tredition®

tredition was established in 2006 by Sandra Latusseck and Soenke Schulz. Based in Hamburg, Germany, tredition offers publishing solutions to authors and publishing houses, combined with worldwide distribution of printed and digital book content. tredition is uniquely positioned to enable authors and publishing houses to create books on their own terms and without conventional manufacturing risks.

For more information please visit: www.tredition.com

TREDITION CLASSICS

This book is part of the TREDITION CLASSICS series. The creators of this series are united by passion for literature and driven by the intention of making all public domain books available in printed format again - worldwide. Most TREDITION CLASSICS titles have been out of print and off the bookstore shelves for decades. At tredition we believe that a great book never goes out of style and that its value is eternal. Several mostly non-profit literature projects provide content to tredition. To support their good work, tredition donates a portion of the proceeds from each sold copy. As a reader of a TREDITION CLASSICS book, you support our mission to save many of the amazing works of world literature from oblivion. See all available books at www.tredition.com.

 Project Gutenberg

The content for this book has been graciously provided by Project Gutenberg. Project Gutenberg is a non-profit organization founded by Michael Hart in 1971 at the University of Illinois. The mission of Project Gutenberg is simple: To encourage the creation and distribution of eBooks. Project Gutenberg is the first and largest collection of public domain eBooks.

Woodward's Graperies and Horticultural Buildings

George E. (George Evertson) Woodward

Imprint

This book is part of TREDITION CLASSICS

Author: George E. (George Evertson) Woodward
Cover design: Buchgut, Berlin – Germany

Publisher: tredition GmbH, Hamburg - Germany
ISBN: 978-3-8472-1533-2

www.tredition.com
www.tredition.de

Copyright:
The content of this book is sourced from the public domain.

The intention of the TREDITION CLASSICS series is to make world literature in the public domain available in printed format. Literary enthusiasts and organizations, such as Project Gutenberg, worldwide have scanned and digitally edited the original texts. tredition has subsequently formatted and redesigned the content into a modern reading layout. Therefore, we cannot guarantee the exact reproduction of the original format of a particular historic edition. Please also note that no modifications have been made to the spelling, therefore it may differ from the orthography used today.

[Pg 1]

WOODWARD'S GRAPERIES

AND

Horticultural Buildings,

BY

GEO. E. & F. W. WOODWARD,

ARCHITECTS & HORTICULTURISTS.

NEW YORK:
GEO. E. WOODWARD & CO., 31 BROAD STREET

[Pg 2]

ORANGE JUDD COMPANY, 245 BROADWAY.
Entered according to Act of Congress, in the year 1865, by
GEO. E. & F. W. WOODWARD,
In the Clerk's Office of the District Court of the United States,
for the Southern District of New York.

[Pg 6]
[Pg 7]

WOODWARD'S

Graperies and Horticultural Buildings.

INTRODUCTION.

It is less than twenty-five years since the first cold Grapery was erected on the Hudson. Since the success of the culture of the delicious varieties of the exotic Grape has been demonstrated, the number of graperies has annually increased, and during the last ten years in a very rapid ratio, until they have become recognized as possible and desirable, among those even whose circumstances are moderate and limited. The newly-awakened interest in this branch of culture is manifested in the number and variety of books and other publications on this subject, the space devoted to it in the agricultural and horticultural journals, and especially in the increased number of graperies and vineyards which have been erected and planted in the last decade. There seems to be a general consciousness of the fact that, in the struggle for wealth and the greed for wide possessions, as well as [Pg 8] in the inherent difficulties of our situation—thrown as we have been upon a new and vast continent—we have too long neglected the culture of the Vine, one of the most ancient and useful arts of life; an art which has, in all ages, been the fruitful source of comfort and luxury, of health and happiness, to the masses of mankind. The neglect of this important and beautiful department of culture is the more remarkable, since our country embraces every degree of latitude, and every variety of climate and soil in which the grape is known to flourish.

It having been demonstrated by years of experiment, resulting in every case in utter failure, that the foreign grape cannot be successfully grown in the open air in the United States—the States of the Pacific excepted—we are obliged to confine our culture to glazed structures, erected for the purpose, where an atmosphere similar to the vine-growing regions of Europe can be maintained, and that bane of the foreign grape, the mildew, avoided.

The culture of choice foreign grapes under glass in this country dates from before the War of Independence, from which time to this the beautiful but perishable Chasselas, the delicious Frontignac, and the luscious Hamburg, have been, here and there, carefully cultivated and ripened. But these efforts have been chiefly confined to the vicinity of large cities, and the [Pg 9] management has mainly been kept in the hands of foreign gardeners, who have imported themselves from the vine regions of Europe, to instruct us in the arts and mysteries of grape-growing.

That many of these are men of great practical experience in the art, we know full well; but, however skillful they may have been in foreign countries, their success in our climate has been achieved only by discarding many of their preconceived ideas, and adapting their practice to agree with the peculiarities of our climate. When the public shall have learned that the culture of grapes under glass is only a plain and simple pursuit or pastime, which any one of ordinary capacity can comprehend and successfully carry out, then we shall have made a decided and important advance.

The American people are rather disposed to be self-reliant, and we may, therefore, safely predict that, when we take hold, in real earnest, of the business of grape culture, either under glass or in the open air, we shall do it with our customary determination and energy, and that success will just as surely follow as it has in other cases where imported ideas have been improved upon and superseded. We have shown, we think, in other fields of enterprise, that we may venture to rely upon native-born talent, ingenuity and industry, to work out this problem also, and that, by [Pg 10] a practical demonstration, we shall, gradually and surely, reach a point of success beyond what has been attained with all the advantages of foreign aid. And this success will be equalled by the simplicity of its methods. Grape-growing in this country is yet in its infancy, and as respects the varieties best adapted to our soil and climate, essentially experimental. As yet it has attracted any considerable attention only of the more intelligent and far-seeing portion of our population, but it is surely beginning to command the regard and study of the larger number of our cultivators, and the inevitable result will be that, in a few years, it must be an important source of our country's wealth.

The great obstacles among us to grape-growing under glass, especially to persons of moderate or limited means, are the first cost of building, planting, &c., and the necessity of regular and systematic care and attention to the vines which must be given, during a short season however, in order to insure success. To those who are influenced by the consideration of such obstacles as these, it may be said that, even in these times of high prices for all descriptions of labor and material—if we except, perhaps, brain-work and intellectual material—complete and substantial grape-houses can be erected at moderate cost, and with proper management they can be made a source of in [Pg 11] come and profit. As to the care and attention required, and the regularity of the periods at which they must be bestowed, at the risk of losing the crop, it can be easily demonstrated that these attentions and duties can be perfectly comprehended and understood by several members of the family, by the older children, and intelligent servants, so as to be overseen and performed by one or another in the absence of the person to whom the care is usually confided. Moreover, when one becomes interested in the management of a grapery, the employment gets to be too fascinating to allow of the thought of restricted action or irksome labor. It soon comes to be regarded as a delightful as well as healthful employment, whose duties are simple, and easily understood and performed.

The love of flowers is becoming quite a passion with many at the present day. This is indicated by the multiplication of nurserymen, and the rapid increase of their sales. Fifteen years ago the sales of flowering plants were confined to a few city Florists; now the trade has become so extensive, that large numbers are grown in our surrounding suburban towns, to meet the demand, which at particular seasons, as the Christmas and Easter holidays, for the decoration of our churches and other purposes, reaches proportions that would surprise the uninitiated. One cultivator has stated that during the fall of 1863 and winter of 1864 [Pg 12] he cut and sent from his establishment, 230,000 blooms of the various flowers he cultivates, and he is but one of many engaged in the cultivation of flowers for the bouquet makers of New York. An extensive grower of pot plants, from information carefully gathered among his fellow nurserymen, estimates that the plant trade of the vicinity of New York reaches nearly the sum of $200,000 annually, and this for plants mainly employed as "bedding plants," in the decoration of gardens and city yards, leaving entirely out of the question, those for winter culture at windows and in green houses, as well as the immense stock of the growers themselves to supply the demand for cut flowers. The growing taste for flowers may be observed in the constantly increasing demand for decorative purposes, in our churches, at public festivals, and private gatherings, and is especially apparent in the numerous depots for their sale on our principal thoroughfares. Much of this is due to the general diffusion of Horticultural literature, unveiling the mysteries of plant culture, and demonstrating the simplicity of the process.

Small green-houses or conservatories attached to dwellings are now frequently to be met with both in city and country: these are entered from some one of the principal rooms of the house, and are an attractive feature both within and without. [Pg 13]

The pleasure derived from such a source is a constantly increasing one, which can only be estimated by those who may have the means for its gratification. But little time and attention is needed, which, with a proper acquaintance with the wants of the various plants, and some experience in their cultivation (knowledge easily and quickly acquired by those who have a genuine love for it), will enable us at any time during the winter season to enjoy our flowers, send a bouquet to a friend, or make use of them in adding to the attractions of home. Such glass structures would afford pleasure to the ladies of the family, in their moments of leisure, being of easy access from the dwelling, without the necessity of exposure to the outer air, which would prevent visits to larger buildings, remote from the house, and could be managed, with occasional assistance in potting and arrangement, wholly by them. Designs for houses of the above character will be found in the course of the work, as well as those adapted as isolated buildings, to grounds of moderate and large extent.

In the construction of Horticultural buildings, the matter of economy is an important and desirable consideration with many persons. But it should be understood that a common, low-priced structure is not the best economy, or the most desirable for a series of years. The dilapidated appearance that soon over- [Pg 14] takes cheap, make-shift constructions, creates an impression that cannot be pleasing either to the spectator or the proprietor. It is an excellent rule, that what is worth doing at all, is worth doing well; and it is just as applicable to horticultural buildings as to any undertaking in life. Rough hemlock lumber, rudely put up and whitewashed, would be a cheap mode of construction, which might be tolerated on a merely commercial place, but would illy correspond with neatly-kept private grounds, however humble and unpretentious they might be. The plan selected may be devoid of mere ornament, which would increase the cost, without adding to the capacity or usefulness, but the proportions should be satisfactory, the arrangement convenient, the materials the very best of their kind, and the workmanship well and faithfully performed. Rough work, open joints, ill-fitting ventilators, ill-proportioned plans and forms, and a general tumble-down appearance, is not the kind of economy we should recommend to our readers or practice on our own place. One may choose between wood and masonry for the foundation walls; between the several grades and sizes of glass; between elaborate finish and ornament, and plain work; in the matter of the various modes of heating, &c.; but whatever is decided upon, let the plan and proportions be correct, and the materials and work of good, honest description. [Pg 15]

In the various designs which we present our readers in this volume, nearly all of which have been erected under our superintendence, and are now in operation, the manner of construction can be judiciously economical, or it may be elaborated to the most substantial and ornamental structures of the class to which they belong. There is no more reason for making these buildings of a temporary character, than there is for putting up our barns and other outbuildings in a cheap and unworkmanlike manner. The enjoyment of a country place naturally depends very much on its neat and tasteful appearance, the completeness of all its appointments, the order and good taste of all its arrangements. And although we do not advocate extravagance, or needless cost in ornamentation, which would be unsuitable to the purpose for which these structures are designed, we think that true economy would indicate the use of the best materials and workmanship requisite for substantial and permanent buildings. Horticultural buildings are not intended for a few years' use merely. Their profit, and the enjoyment they afford, will last for many years, and may be transmitted, with the other improvements of the country seat, as substantial and attractive appendages, indeed, as real property, worth all the money they cost, to the future proprietor.

There is still much to be learned in the matter of [Pg 16] exotic grape-growing in this country, and, in fact, in the management of conservatories, orchard-houses, and all descriptions of horticultural buildings, and all classes of plants cultivated under glass. Whatever progress may have been made abroad, where experiments are carried on upon a large and costly scale, and often with eminent success, is of little or no value to the American horticulturist. Our climate is very different in its character and conditions from that of Europe, and especially that of humid England. We have, what they lack, real sunshine, with clear skies. Under the English methods of treatment, our graperies and green-houses would speedily be ruined. Nor are we willing to accept as final and conclusive the present best-known methods of vine culture. If there are better modes of managing exotic or native vines, and of developing the whole theory of grape culture, we shall be quite sure to find them out in the wide sweep of experiment which we are boldly and patiently undertaking in various parts of the country.

We do not propose, in our present work, to enter upon the investigation and discussion of the various theories of heat, light, color, radiation, &c., which properly belong to scientific treatises on these subjects. We intend to give only practical examples and results, from an extensive professional experience, with numerous designs and plans of buildings, most of which [Pg 17] are now in successful operation, with the expectation that this volume will contribute not only to the general information of our horticulturists, and of gentlemen who are establishing themselves in the country, but also to create and encourage a taste for this kind of culture of exotic and delicate fruits, as well as the exquisite but tender gems of the floral world. When we find that we can command, at comparatively small cost of money and attention, the beautiful and luscious fruits of southern and tropical climes—their rarest and choicest flowers—the most delicious grapes, the finest peaches, nectarines, and apricots, the fig, and the pineapple, if we will; and that we can command these in abundance, to load and adorn our tables daily, the time cannot be distant when horticultural buildings, of various descriptions, will be found on all our country places or attached to our city homes.

POSITION OF HOUSES.

For lean-to or single-roofed structures used as forcing-houses for grapes or other fruits or plants, a southern aspect is generally preferred. Our own preference would be a position facing South-East, on account of the advantage gained from the morning sun, which is so favorable to the health and growth [Pg 18] of all descriptions of plants. Although an hour or two of the evening sun might be lost to a building in this position, yet the rays are then comparatively feeble, and this loss would be much more than compensated by the more genial morning light.

Cold Graperies, with span roofs, and glazed at both ends, are better placed North and South,—that is, with the ends facing these points,—as nearly as a due regard to the positions of other buildings in the vicinity, and the general symmetry and apportionment of the grounds will permit. Each side of the roof will thus receive an equal amount of sun-light. For span-roofed Green-houses the rule is not so arbitrary, the glass not being lined with foliage, as in the case of graperies, the diffusion of light would not be materially obstructed. Under some circumstances, Green-houses may be placed east and west, as when a portion of the house is to be devoted to the purposes of propagation. The north side can thus be advantageously used, being less exposed to the sun's rays. Many plants requiring partial shade, would find there, also, the most favorable conditions for their cultivation.

Green-houses or Conservatories attached to dwellings, will answer in almost any position that convenience may require, or the taste suggest, as they are generally not so much intended for the growth of plants as for their display when in bloom. The sun [Pg 19] should shine upon them, however, at least half the day. When they are intended for the growth of plants, then the more sun-light they can have the better.

FORMS OF HOUSES.

Until within a few years past, the straight-pitched roof, both single and double, has been used almost exclusively in the construction of glass houses. That there is an advantage in this form over some others, on the score of expense, and because there is less skill required in the builder, we admit, but there the advantage ends. The superiority of the curvilinear form is now beginning to be very generally acknowledged, on account of its being more graceful and pleasing to the eye, and because of its superior adaptability to the growth of plants. When to the curved roof is added the further improvement of circular ends, as illustrated in some of the designs furnished in this work, we have secured forms of houses that will admit double the light of the old-fashioned, heavy sliding sash structures which were built twenty-five years [Pg 20] ago. Happily these old glass houses are fast falling into decay, and but few new ones are erected on their model.

Curvilinear roofs possess advantages over those of a straight pitch which may be briefly summed up as follows:

1. A larger run of roof for a given width of house, and consequently, more and better diffusion of light.

2. A greater power of reflecting the sun's rays, because of the constantly varying angle at which they strike the glass.

3. A greater amount of head room within the building, without the necessity of high parapet walls, or perpendicular sides.

4. Greater strength of the roof, enabling it to resist pressure from accumulated snows, without the necessity of supporting columns under the rafters, which are indispensible under a straight roof of considerable span, to prevent its settling down, and the opening of joints in glass and wood work, admitting the cold air from without.

A good proportion for a grapery or conservatory, is twenty feet in width by fifty feet in length. We think the width should never be much less where the roof is of double pitch. Single pitched houses should not exceed sixteen feet in width. [Pg 21]

Mistakes are frequently made in the erection of structures for the growth of plants which, notwithstanding all the skill and art of experienced gardeners, render it impossible to arrive at satisfactory results. One of the most common of these is the excessive height of the roof. Men of experience in the construction and use of glass houses, have satisfied themselves that the lowest elevation which the uses and purposes of the building will admit, is the best. The difference in temperature between the floor and roof of a house twenty feet in height, will vary from ten to fifteen degrees. It is obviously desirable that there should be as little difference as possible in the temperature of the air on the ground, among the lower parts of the plants, and in the upper regions of the house. The nearer we can approach an equilibrium, the better success will attend our efforts. Nurserymen generally, and sometimes other cultivators, understand this, and they build their plant houses with roofs of low pitch, affording scarcely room to stand upright within them. Their plants are thus brought near the glass, and they grow stocky and firm, presenting quite a different appearance from the attenuated specimens frequently met with in private establishments. [Pg 22]

HEATING.

The proper heating of Horticultural buildings being an important feature in their general management, and an essential condition of their success, we shall consider the subject at some length, availing ourselves of the practical experience of others, as well as of the knowledge we have acquired in our own experiments and practice.

Hot air stoves have been so generally condemned and discarded as a means of heating glass structures, that we shall not discuss their faults or merits, but confine ourselves to heating by flues, steam, and hot water in pipes and tanks.

Flues.—Flues have been generally used in heating for many years, and although the method is rude, imperfect and unsatisfactory, they possess certain advantages on the score of economy, which will prevent their total supercedure until some equally cheap and effective method shall be found, to take their place. It cannot be questioned that houses of moderate extent can be heated at much less expense for the original cost of apparatus by the flue system than by any other now before the public. Flues have the advantage over steam or hot water in their power to generate heat and supply it to the green or hot house [Pg 23] in a very short space of time, and with this apparatus, the fires may be allowed to go out on mild and bright days in winter, with the certainty that heat can be easily and quickly commanded at nightfall. Steam cannot be generated quickly, and the hot water apparatus requires considerable time to get into full operation, with the usual amount of fuel.

Among the serious objections to the use of flues, is the unequal distribution of heat throughout the house; the parts near the furnace being overheated, while at the chimney it is scarcely warm. This difficulty can be partially obviated by the use of materials in the construction of the flues, of different thicknesses,—being made thick and heavy at the furnace, and gradually becoming thinner and lighter as it extends towards the chimney. Again, flues generally require more fuel than a hot water apparatus, and moreover, they are unsightly in an ornamental house, and with the best care in their construction and management, they do not give entirely satisfactory results.

Earthenware drain-pipe is frequently employed for flues, and when care is taken to prevent their cracking by the excessive heat near the furnace, they answer the purpose very well. When properly secured at their joints they prevent the escape of gaseous matter more perfectly than brick flues.

Flues should be elevated a few inches above the [Pg 24] floor, and supported by bricks, to allow all the radiating surface to act upon the atmosphere of the house, and should have, in order to secure sufficient draft, a gradual rise through their whole length from the furnace to the entrance into the chimney.

The furnace should be built inside the house at one end, with the fire and ash-pit doors opening into a shed outside, to prevent any escape of gas into the house while replenishing the fire. It will be necessary to place the furnace low enough to allow a proper rise to the flue. If the flue be made to rise immediately from the furnace about one foot, it may then be carried fifty feet, with a rise of not more than six inches, and the draft will then be sufficient.

The dimensions of the flue may vary from 8 to 12 inches in width, and from 12 to 18 inches in height, according to the space required to be heated. The usual mode of construction, when bricks are used, is to lay them crosswise and flat for the bottom and top, and to set them edgewise for the sides. Tiles for the bottom and covering are an improvement upon bricks: being thinner, the heat passes through them more readily, while they still retain the heat sufficiently to equalize the temperature. Tiles used for the top covering are sometimes made with circular depressions for holding water for evaporation. [Pg 25]

Steam.—The employment of steam for heating green houses, graperies, &c., is almost entirely superceded by the hot water method. It will, therefore, be necessary only to allude briefly to this part of our subject. It occasionally happens that a conservatory attached to a dwelling is heated by the same steam apparatus employed to heat the latter, but we believe that a person who should advocate, at the present day, the general adoption of steam as a means of heating horticultural structures, would be regarded as belonging to a generation which has now passed away.

Steam travels through pipes with great rapidity, and parting with its heat rapidly, it becomes quickly condensed, unless the boiler is of large capacity and capable of furnishing a full supply. It is, at best, an unsatisfactory mode of heating plant houses, for if from any cause the water in the boiler is reduced below the boiling point, the steam in the pipes is instantly condensed, and with it all heat, except that remaining in the iron of the pipes, and the condensed steam, is withdrawn.

Hood, an English author on heating, quoted by McIntosh in his valuable work the "Book of the Garden," thus compares the merits of steam and hot water. "The weight of steam at the temperature of 212° compared with the weight of water at 212°, is about as 1 to 1694, so that a pipe that is filled with [Pg 26] water at 212°, contains 1694 times as much *matter* as one of equal size filled with steam. If the source of heat be withdrawn from the steam pipes, the temperature will soon fall below 212° and the steam immediately in contact with the pipes will condense: but in condensing, the steam parts with its *latent heat* and this heat in passing from the latent to the sensible state, will again raise the temperature of pipes. But as soon as they are a second time cooled down below 212° a further portion of steam will condense, and a further quantity of latent heat will pass into the state of heat of temperature, and so on until the whole quantity of latent heat has been abstracted and the whole of the steam condensed, in which state it will possess just as much heating power as a similar bulk of water at the like temperature; that is, the same as a quantity of water occupying 1-1694th part of the space that the steam originally did.

By experiments made by the above authority, it has been proved that a given bulk of steam will lose as much of its heat in one minute as the same bulk of hot water would in three hours and three quarters. And further admitting that the heat of cast iron is nearly the same as that of water, if two pipes of the the same calibre and thickness be filled, the one with water and the other with steam each at 212° of temperature, the former will contain 4.68 times as much [Pg 27] heat as the latter; therefore if the steam pipe cools down to 60° in one hour, the water pipe will take four hours and a half to cool down to the same point. In a hot water apparatus we have in addition to the above, the heat from the water in the boiler, and of the heated material in and about the furnace, which continues to give out heat for a long time after the fire is totally extinguished; whereas in a steam apparatus, under the same circumstances we have no source of heat except the pipes by which it is conveyed—giving an advantage in favor of hot water over steam as regards its power of heating hot houses, and maintaining heat after the fire ceased to burn, in nearly the proportion of 1 to 7—that is, hot water will circulate from six to eight times longer than steam under the above circumstances."

Tanks.—This mode of heating horticultural buildings has been used in England for some years, and has, of late, obtained considerable popularity in this country; mainly, however, for the purpose of obtaining bottom heat. The tank method is more steady and reliable in its operations in this respect, than heating by flues or pipes, but even its most strenuous advocates must admit that for atmospheric heat hot water pipes or flues must be employed in some shape or other, where the tanks are covered with earth or [Pg 28] sand beds for propagating purposes. With slate or metallic covering they are sometimes used solely for atmospheric heat, and are found to answer well. But if tanks are constructed of substantial and enduring materials, they possess little if any advantage, on the score of expense, over hot water pipes, while they occupy much more room and are unsightly objects in a well ordered green-house.

Wooden tanks are frequently used where the heat is required to rise perpendicularly from them. If constructed of good pine plank, well put together with white lead, and thoroughly painted inside and out, they will last for several years. Scarcely any heat will be radiated from the sides and bottom of a wooden tank. Tanks of brick and cement would answer better than those made of wood, if it were possible to make them water-tight when supported by piers above the ground, as they are usually built. But however carefully constructed, these materials are so unyielding to the expansion and contraction they are subjected to, that it is nearly impossible to prevent leakage for any length of time. A large number of brick and cement tanks have come under our notice, and we cannot call to mind a single one of them all that has not been a continual source of vexation and expense to its owner, since its first construction. [Pg 29]

The principle objections to tank heating, as usually employed, are an excess of bottom heat and a deficiency of atmospheric heat, with a superabundance of moisture when the vapor from the tank is not properly excluded from the house. Tanks should be covered with some good radiating material, as slate or metal. If slate is employed, the joints should be carefully and effectually cemented. Boards are sometimes used as a covering, but their radiating power is slight, and their decay rapid.

Soil or sand, to the depth of six to ten inches, is usually placed upon the tanks, and used as a plunging bed for pots containing cuttings; or the cuttings are sometimes inserted in the bed itself.

Any arrangement by which vapor from the tanks is admitted to the roots of plants is to be avoided, for however desirable a moist bottom heat may be, it is found from experience that the soil is frequently rendered a mass of puddle, in which no living roots can exist.

A portion of the covering of the tank may be made moveable to allow moisture to escape into the house when required.

By means of the tank, bottom heat for propagating or other purposes, can be very steadily and uniformly maintained, more so than by other modes, and the changes of temperature of the outer air do not ma [Pg 30] terially affect it. But the case is different with regard to the air of the house, which is frequently reduced below the freezing point, in severe weather. If the bottom heat is of the required temperature, any attempt to counteract the coldness of the air of the house by increasing the fire, would produce an injurious excess of bottom heat. It is evident that while the required supply of heat for the bottom is uniform, and that for the top exceedingly irregular, both objects cannot be properly secured except by a separate supply of heat for each. For these reasons we would employ a hot water pipe or pipes, passing around the house, on the same level with the tanks, supplied with a valve to regulate the heat at pleasure, or a brick smoke flue constructed in the usual manner.

Tanks are usually divided in the centre, thus forming channels for the flow and return circulation side by side, equalizing the temperature throughout their whole length. This form is sometimes departed from by carrying the tank around the house, and connecting each end with the boiler, but in this case, except in small houses, a uniform temperature cannot be maintained, as the water will have lost several degrees of heat before it has accomplished its circuit. Another arrangement is to connect the remote end of the tank by an iron pipe for the return circulation, passing under the tank the whole distance to the boiler. This [Pg 31] is not as perfect and effective an arrangement of pipes and tanks as that before referred to, as in this case we do not have the heat from the pipe under control.

A writer in a late number of the "Gardeners' Monthly," gives the following description of tanks erected by him to obviate excessive moisture and radiate a portion of their heat into the atmosphere of the house.

"In the winter of 1863-4, I finished two span-roof houses, each 60 feet in length, with water tanks three feet in width, running entirely around on both sides of each house, and heated by a single furnace. The tanks were made with wooden bottoms and sides, and covered with slate carefully cemented. My design was to heat the houses entirely by the tanks, by far the larger portion of the heat being given off from the slate covering, and as a bottom heat for plants. As I understand the various writers upon this subject, this is the approved plan. But I have found considerable difficulty, and have been obliged to modify my plan in various respects:

In the first place, wooden tanks, with the top covered with sand, will not give off heat sufficiently to keep up growth in houses of this size during extremely cold weather. By protecting the houses with shutters, this difficulty may be obviated. Crowding the fire, and raising the water in the tanks to a high [Pg 32] temperature, is a more objectionable remedy. In this way the bottom heat is too strong. But my most serious difficulty has arisen from excessive humidity. I put three inches of sand over the whole slate surface of the tanks, using a part for cuttings, and the rest, (say 100 running feet of the three feet wide table), for standing pot plants upon the surface of the sand. The plants dried rapidly, and required watering every morning. The result was, that in watering the plants, and of course the sand on which they stood, to some extent, it was like pouring water upon a flue, or upon hot pipes: a constant steam was given off; all the moisture in the sand was rapidly converted into steam; so, also the water in the pots was quickly expelled. In order to heat the house sufficiently, the bottom heat became too strong, and the plants were in too direct contact with it. In cold days the house was in a perfect fog. It was ruinous to the plants. The remedy was simple: more heat must be allowed to escape from the tank into the house, without coming in contact with the sand-bed, and the moist earth of the plants. Another slate floor was laid, an inch above the tank slate, on which to put the sand and stand the plants. This hot air chamber opens into the house on the back and front side of the tank. Thus the whole radiating surface of the top of the tank may be directed into the house, or may be con [Pg 33] fined as bottom heat, as may be found necessary. By this plan, excessive humidity may be entirely obviated, and the heat completely controlled, as wanted."

Hot Water Pipes.—It is generally conceded, among practical men, that the circulation of hot water in iron pipes is the best known method of heating plant houses. The property which heated water possesses of retaining for a considerable length of time its heat and transmitting it to the pipes at long distances from the boiler, renders it a most effective agency for such purposes: A perfect control of the moisture of the atmosphere, by means of evaporating pans attached to the pipes; entire freedom from deleterious gases, sometimes escaping from flues, and the substantial character and enduring qualities of the apparatus, are important considerations in favor of this method of heating which are not to be overlooked or underrated.

It is true that a house of a given size cannot as soon be brought to the required temperature after the fire is first lighted, as by other modes of heating, but when once in full operation greater regularity is maintained, and if the fire should by any neglect go out, heat is still radiated, often for several hours, before the pipes become entirely cold.

For heating ornamental houses of glass, pipes are [Pg 34] also to be recommended on account of the little room they occupy and the neatness of their appearance compared with the unsightly flues or tank. If properly put up, the pipes never leak at the joints, as is the case frequently with tanks, and scarcely need any repairs for years. The first cost of apparatus for heating by hot water pipes exceeds that of the other methods which we have named, but when we take into account its great durability, economy of fuel, and the satisfactory results produced in the growth of plants in houses heated in this manner, it must be evident that this method is the cheapest in the end.

It is generally supposed that the heat obtained from steam or hot water pipes necessarily contains moisture. For those who have had any experience in the use of these methods of heating, it is needless to say that such is not the case. To obtain moisture evaporation of water in some manner in the atmosphere must be effected. This is provided for by attaching to the pipes evaporating pans filled with water, by which the moisture can be perfectly regulated and controlled. The capacity of the boiler and the length of the pipes should be in proportion to the size of the house to be heated, bearing in mind that it is better to have a reserve of heating power for extraordinary occasions. In such cases economy in fuel will be [Pg 35] secured, as the fires will not be required to be kept constantly burning brightly.

Fault is sometimes found with the apparatus when it lies entirely with the proprietor of the establishment, who in his short-sighted economy, has restricted the builder in the amount of pipe put into the apparatus.

CONSTRUCTION, &c.

The general plan of Horticultural structures may be as perfect as possible, but if the details are not well carried out, and especially if the workmanship be not good, they will prove a source of never-ending vexation and expense. Insecure foundations, ill-fitting doors and ventilators, imperfect glazing, and inferior workmanship of every description, are evils that skillful gardeners have to contend with, and upon whom the consequences of such defects usually fall, when they should be placed upon the shoulders of the constructor.

Methods for building cheap Graperies and Green houses have often been described, and we find many of these imperfect and temporary structures scattered through the country. Such buildings may be cheap [Pg 36] as respects their first cost, but their durability is a question which should enter into the calculations of their builders, as well as the consideration of the original outlay. After a year or two we find them with open joints, leaky roofs, and decaying foundations. The inferior and temporary character of materials and workmanship is often a source of serious loss to their owners, and every building of this description demonstrates the mistaken and short-sighted economy of its projector. It is much wiser and truer economy to expend at the outset, a sufficient amount of money and care to make the structure permanent, and to obviate the necessity of constant repairs. Experience has taught us that if they are well and substantially built, these structures will endure for twenty years with very few repairs except an occasional coat of paint. It need not be demonstrated that the profit and gratification to be derived from a well-built house far exceed those accruing from a cheap and imperfect one, with escapes for the heat in winter, and inlets for cold air and driving snow and rain.

The foundations of Horticultural buildings should be of stone or brick, both below and above the ground, if they are to be of a permanent character. The superstructure should be of the best white pine and thoroughly painted. In building curvilinear roofs the rafters and sash bars should be sawed out in pieces [Pg 37] to the regular curve. The rafters being put together in sections, breaking joints are thus equally strong throughout their length. The advantages of sawed bars over those bent in the usual manner, are very great. The thrust of the roof is but slight, and the house always remains in shape. With the bent bars the strain is enormous, as may be seen in the settling of such houses at the ridge, and expansion at the sides, besides the liability of breaking the glass by the constantly varying strain of the bars.

Iron has been frequently and strongly recommended in the construction of horticultural buildings. It has been used, with very satisfactory results in England, and doubtless it may there be found to be the best and most economical material for such purposes. It has been tried also in this country, but the experiment has not resulted so favorably. The main difficulty is that, in this climate, the expansion and contraction of the iron rafters and bars are so great that the glass is continually and badly breaking, and it is very difficult to keep the joints tight enough to repel the rain and the cold air. There can be no doubt that in this country, wood is a better material than iron for these purposes.

Thick and double thick glass has heretofore been used almost exclusively for first class houses, but the high price of glass has of late, compelled the use of a [Pg 38] thinner article. It is generally believed that thick glass will resist hail storms better than thin, but on this question practical men differ in their opinions. It is contended, on the other hand, that the elasticity of the thin panes resist a blow better than the unyielding thick one, also that the latter is more likely to be broken by the accumulation of water between the laps of the glass.

We have found that the 8 by 10 size of single thick French window glass, second or third quality, is sufficiently good for Horticultural buildings, and we do not use any other, unless especially called for by the proprietor.

Glazing is often badly executed, half an inch lap, and sometimes more, being often allowed to the glass, from the mistaken idea that rain, in a driving storm, will find its way through. A lap of one-eighth of an inch is amply sufficient in any case. The glass should be well "bedded" down to the sash bar, in putty containing a portion of white lead, and well secured with small iron nails or glaziers points. All putty should be removed from the outside when the work is finished, and the sash bars should then be painted with a heavy coat of thick paint which will close up the joints and render them water tight.

Ample ventilation should be provided both at the top and bottom of houses, so that large quantities of [Pg 39] air may be supplied when necessary, as in ripening the wood of vines in graperies, and in "hardening off" plants in green houses before removal to the open air.

By reference to the numerous designs given in this work, the manner of arranging the interior details, such as shelving, tables, walks, hot water pipes, and the general features of construction and adaptation, will be understood.

HOT-BEDS.

The most simple form of Horticultural structures, and one known in almost every garden, is the Hot-bed. To persons of experience in their construction and management, we cannot hope to give any important information, but having seen in many instances the operations of these beds imperfectly performed, we offer a few simple suggestions and directions which will be of advantage to the novice.

The location of the bed should be, if possible, a sheltered one, especially on the north side, while towards the east and south it should be open. This shelter or protection is needed chiefly to prevent an undue radiation of heat from the glass, and the entrance of a strong, cold current of air when the [Pg 40] sashes are lifted for ventilation. This radiation is not only hurtful to the plants by causing sudden and extreme changes of temperature, but, if allowed to proceed too far, will cause the heat of the bed to "run out." Let the shelter, therefore, be as thorough as possible.

We have found the south side of a barn, or a tight board fence a good location. The barn would be preferable, on account of its proximity to the materials that furnish the source of heat—the manure pile.

If the soil is wet, or of a heavy nature, it would be better that the bed be made entirely upon the surface. If the situation is a dry one, and the soil gravelly or sandy, then a pit may be excavated, of the size of the intended frame, and three feet in depth. A hollow brick wall should be built up from the bottom, six inches above the surface, if it is intended that the bed should be permanent; otherwise the excavation may be lined with boards, or if designed for only a season's use, it may be left without any support. Hot-beds made under ground require less material, are more lasting in their heat, and require less attention than those built on the surface. On the contrary, should the heat fail from any cause, beds built up on the surface possess the advantage of being more easily renewed by the application of fresh fermenting materials, or "linings" as they are usually termed. [Pg 41]

About the 20th of February is early enough, in this latitude, to gather and prepare materials for the hot-bed. Fresh stable manure alone may be used, though preference is generally given to a mixture, in equal proportions, of manure and forest leaves. Place on the ground, (under a shed if possible,) a layer of leaves one foot thick, and on this a foot of manure, then leaves and manure alternately until the required quantity is obtained. Let this heap remain four or five days, or until it begins to heat, then turn over and thoroughly mix the leaves and manure together, and throw them up into a compact, conical heap. In four or five days more your materials will be ready for your bed. Mark off your intended site, running as nearly east and west as practicable. Your frame should be about six feet wide and of any required length. The manure bed should extend a foot outside the frame on the sides and ends. See Figure 1, in which *a* is the manure heap.

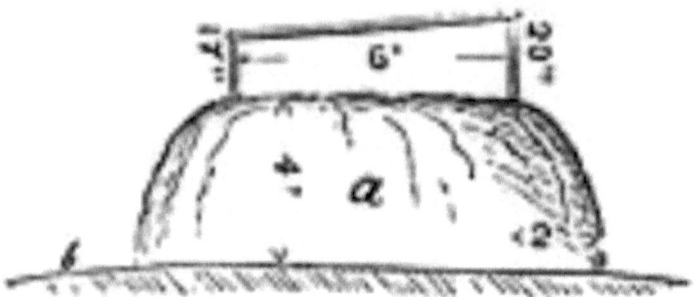

Fig. 1.

[Pg 42]

Build up the manure square and level, shaking, mixing, and beating it with the back of the fork, to the height of about four feet, making the centre somewhat higher than the sides, to allow for settling. The frame should be of 1-1/4 inch pine, twenty inches high at the back, and seventeen inches in front, and may be put together with hooks and staples, so as to be removed and stored, when not in use. The sashes should be six by three and a half feet, and the frame should have cross-bars at every sash for support. It is well to have the frame divided by partitions into two or three compartments, that one section may receive more or less ventilation as the plants grown in them may require. In three or four days the heat will be up in the bed, and then it should be covered with six inches of fine garden mould, which should be raked off level. When the soil is heated through, the seeds may be sown. Ventilation should be given to let off the steam and vitiated air, but with caution to avoid the loss of heat. Straw mats will be required to cover the sashes at night, and should be regularly put on. If the weather is very cold, shutters or boards in addition are necessary. If care is exercised in the management, the heat will be maintained as long as is desirable.

Figure 2 represents the hot-bed partly beneath the surface. [Pg 43]

The frame in this case will be fifteen inches in height at the back, and twelve inches in front, constructed in the same manner as that before described. The materials and the general preparation of the bed is also the same. A space of about eight inches should be left between the surface of the mould and the glass, to allow for the growth of plants before the sashes can be removed. Coarse litter should be put around the frame, and up even with the top of it, to confine the heat. Beds should be well covered before the sun has left them in the afternoon, and not opened in the morning until the sun is well up. Seeds of vegetables for early planting, and those of annual flowers may be sown, and cuttings of green-house and bedding plants started in pots. Such a bed will also be a favorable place for the propagation of grape eyes, in which an experienced person will often succeed better by this humble means, than with the best designed and most conveniently arranged propagating house.

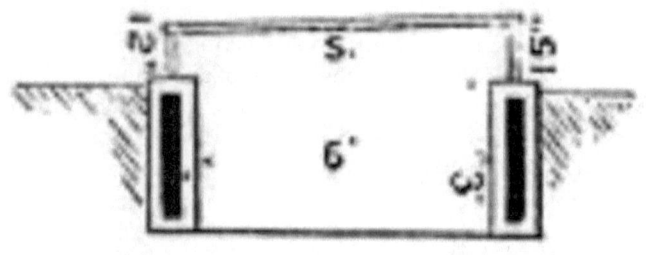

Fig. 2.

[Pg 44]

THE COLD PIT.

Many who have not the advantage of a green-house, wish to preserve over the winter their half-hardy plants which have ornamented their garden during the summer. These are generally consigned to the cellar to dry up and be forgotten. In the darkness they loose their leaves, and when in spring they are again brought to light many are dried up and dead. Properly constructed cold pits offer superior advantages for the protection of many plants of a half-hardy nature, and indeed some that are usually considered tender here find a congenial location. Such a pit should be permanent in its character, and located in a spot easy of access to the house, that it may receive proper attention during the winter. A convenient size, and one sufficient for an ordinary garden would be ten feet long by five wide, varied somewhat from these dimensions to suit size of glass in sashes. The pit should be excavated four feet and a half below the surface, and a hollow wall of brick built up to one foot above the surface. Six inches in depth of coarse gravel should be placed in the bottom on which the pots containing the plants rest. Shelves may be also placed around the sides for the smaller plants. The wall above the ground should be "banked up" to within three inches of the top and sodded. [Pg 45]

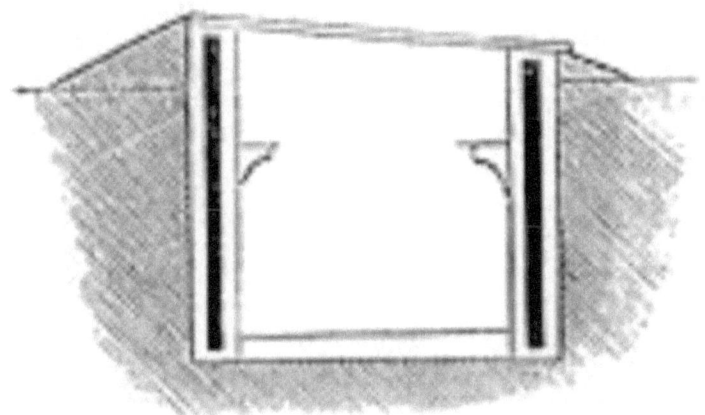

Fig. 3.—Cold Pit.

Double sashes we have found give great protection and save attention in covering the pit. The bars of these sashes are "rabbited" on both sides and double glazed, thus enclosing a stratum of air affording a good non-conductor of heat from within, or cold from without the pit. The plants when first put in the pit will require to be watered and the sashes opened during the day, until cold weather. But little water is required during winter, as the plants are in a state of rest, and partial dryness at the roots is of advantage. In very severe weather straw mats would be required, but the double glass would keep out 10 to 15 degrees of frost. Some ventilation must be given on mild days when the sun is bright to carry off the dampness, but in dull cold weather all should be kept closed up. Camellias and Azaleas do admirably in such quarters, and can be brought into the dwelling [Pg 46] and flowered at any time during the winter. Many plants grow with surprising luxuriance after remaining dormant in such quarters all winter. As the season advances in the spring ventilation must be given during the day, closing the sashes at night until the weather becomes mild when they may be gradually removed altogether.

PROPAGATING HOUSES.

Cheap and effective propagating and plant houses, for Nurserymen, have become of late years a necessity from the great increase of the trade in flowering plants for the decoration of our gardens and green-houses, and the very extensive demand for the new and superior varieties of the native grape. Peter Henderson, Esq., of Jersey City, long known as an extensive and successful propagator, in an article written for the Horticulturist, thus speaks of his house and management:

"After many years of extensive practice, I have arrived at the conclusion that cuttings of almost every plant cultivated by the florist or nurseryman will readily and uniformly root, if the proper conditions of temperature and moisture are given them. It [Pg 47] matters little or nothing how the cutting is made, or what may be the color or texture of the sand or soil in which it is planted; these have little or nothing to do with the formation of roots. But an absolute condition of *invariable* success is uniformity of temperature and moisture. To attain this uniformity, the structure of the house is of vital importance; and it is owing to the erroneous construction of buildings for this purpose that so many have to deplore their want of success. I will briefly describe the construction of the propagating pit we have in use, and the manner of operations, which will best explain my views on the matter. The pit, which faces north, is 65 feet in length by 8 in width, and 3 feet high at back by 1 in front, the pathway being dug out to give head-room in walking. The front bench is 3 feet wide, walk 2 feet, and back bench 3 feet. All along the front bench run two wooden gutters 9 inches wide by 3 inches deep, the water in which is heated by a small conical boiler connected by two pieces of leaden pipe to the gutters. Three inches above the water in the gutters is placed the slate or flagging, (resting on cross slats of wood,) on which is two inches of sand. By regular firing we keep a temperature *in the sand* from 55 to 75°; and as the pit has no other means of heating, except that given out by the sand in the bench, the atmosphere of the house at night is only [Pg 48] from 40° to 50°, or 25 degrees less than the "bottom heat." In the daytime, (in order as much as possible to keep up this disparity between the "top" and "bottom" heat,) a little air is given, and shading the glass resorted to, to enable us to keep the temperature of the house down. And here let me remark, that when propagation is attempted in green-houses used for growing plants, (such houses facing south or southeast,) the place usually used for the cuttings is the front table; and it being injurious to the plants to shade the whole house, that part over the cuttings alone is shaded; the consequence is, that the sun, acting on the glass, runs the temperature of the house up, perhaps, to 80°, or *above* that of the bottom heat, the cuttings wilt, and the process of rooting is delayed, if not entirely defeated. All gardeners know the difficulty of rooting cuttings as warm weather comes on. When the thermometer marks 80° in the shade fires are laid aside; and if the rooting of cuttings is attempted, the sand or soil in which they are planted will be 10 or 15 degrees *lower* than the atmosphere, or the opposite of the condition required for success.

The advantage possessed by the gutter or tank, as a means of bottom heat, over smoke flues or pipes, is in its giving a uniform moisture, cuttings scarcely ever requiring water after being first put in, and then only to settle the sand about them. Still, when this [Pg 49] convenience is not to be had, very good success may be attained by closing in the flue or pipes, regularity in watering, and a rigid adherence to these degrees of temperature.

The propagating pit above described is used for the propagation of all kinds of plants grown by florists, such as Camellias, Dahlias, Roses, Verbenas, Fuchsias, Grape Vines, etc. The time required in rooting cuttings of soft or young wood is from seven to ten days. Last season, during the month of February, we took three crops of cuttings from it, numbering in the aggregate forty thousand plants, without a loss of more than one per cent. In fact, by this system we are now so confident of success, that only the number of cuttings are put in corresponding with the number of plants wanted, every cutting put in becoming a plant.

In this narrative of our system of propagating, Mr. Editor, I have not attempted to theorize. I give the plain statement of operations as we practice them, thoroughly believing that the want of success in every case must be owing to a deviation from these rules. Ignoring entirely most of the maxims laid down in the books, such as "use a sharp knife," and "cut at a joint," we use scissors mostly in lieu of a knife, and we never look for a joint, unless it happens to come in the way. We are equally skeptical as to the [Pg 50] merits of favorite kinds and colors of sands or other compounds used for the purpose. Of this we have reason to be thankful, for a nicety of knowledge in this particular in the head of a scientific (?) propagator may sometimes become an expensive affair.

A friend of mine, a nurseryman from the far west, deeply impressed with our superior horticultural attainments in the Empire City, hired a propagator at a handsome salary, and duly installed him in his green-house department; but, alas! all his hopes were blighted. John failed — signally failed — to strike a single cutting; and on looking about him for the cause, quickly discovered that the fault lay entirely in the sand! but my gullible friend, to leave no stone unturned, freighted at once two tons of silver sand from New York to Illinois! Need I tell the result, or that John was soon returned to where the sand came from?"

During the past year, Mr. Henderson has erected an extensive range of houses, after the following description and plan:

"I have read and examined from time to time, with much interest, your remarks and sketches of Plant Houses, and it is not to dissent from your views that I now write, although it seems to me that your ideas run all one side of the matter, for your designs and descriptions are almost exclusively of an ornamental [Pg 51] character, and adapted only for conservatories or graperies, leaving the uninitiated commercial nurseryman or florist to look in vain for something to suit his case. I have said that your ideas seem to be one-sided, in describing only ornamental erections; they seem also so in your uniformly recommending the fixed roof principle. Now, for the purposes of the florist or nurseryman, I think there is but little doubt that the advantage is with the sash over the fixed roof. The difference in cost is trifling; probably a little in favor of the fixed roof; but balanced against that is, that your house, once erected on your favorite plan, you are emphatically "fixed." It is not portable, (unless made in sections, which is only a bad compromise with the sash plan,) and any alteration requiring to be made, your roof is of but little or no value. But the most serious objection to it is the difficulty with air. I have never yet seen a house built on the fixed roof principle that had means of giving air so that plants could be grown in a proper manner, and I could name dozens who have been induced to build on this plan, that one year's experience has given them much reason to regret.

Fig. 4. *a*, ground level. – *b*, bench or table on which to stand plants, 4-1/2 feet wide. – *c*, 4 inch pipe, 3 in each house. – *d*, pathway, 2 feet wide.

We are now adopting for plant houses, low, narrow, span-roofed buildings, formed by 6 feet sashes, one on each side, the *ends* of the houses facing north and south. These we attach three together, on the "ridge [Pg 52] and furrow" system, as shown in sketch. This system presents great advantages, and, by using no cap on the ridge piece, air is given in the simplest and safest manner, by the sash being raised by an iron bar 9 or 10 inches long, pierced with holes, which answers the double purpose of giving air and securing the sash, when closed, from being blown off by heavy winds. There is no necessity for the sashes being [Pg 53] hinged at the bottom, as might be supposed; all that is required being to nail a cleet along the wall plate, fitted tight to the bottom of each sash. Every alternate sash is nailed down; the other is used in giving air in the manner described.

The advantages of such erections are so obvious, that I need not trespass much on your space to enumerate them. The plan can be adapted to detached buildings already up, by erecting houses of the same length alongside; or, in the erection of new houses, if not more than one is wanted, it may be put up with a view to further extensions. I have had four houses on this plan in operation for nearly two years, and I have never before had so much satisfaction with any thing of the kind. Intending next season to remove my greenhouses from their present site, all shall be put up after this style."

Messrs. Parsons & Co., of Flushing have also built several houses similar in design for the propagation of [Pg 54] grape vines. These latter are heated by brick flues and have proved very satisfactory. The vines are grown in beds and not staked. Pot culture in the usual manner would require greater height of roof. The only objection that we can see to houses built in this manner is the accumulation of snow in the furrows. Mr. Henderson assures us that this is not an objection of any moment in this latitude, and that the expense attending the removal of snow is too slight to be considered.

DESIGN No. 1.

Figures 5 and 6 are a section and ground plan of a propagating house for growing grape vines, but it might serve as well for other plants. The length of the house is on an east and west line, giving a northern exposure to the roof on one side, the opposite facing the south. A board partition runs through the centre dividing the house into two. This partition might be made movable, so that at any time the house could all be thrown into one. The foundations are of stone projecting 6 inches above the ground. Two and a half feet of vertical boarding, above which is two feet of sash, give a height of four and a half feet above [Pg 55] the foundation for the side of the house. The side sashes are hinged for ventilation. Top ventilation is afforded at the ridge by ventilators raised by rods from the inside. The roof is on the fixed principle that is composed of sash bars extending from plate to ridge, in which the glass is set. In the north division a combination of the tank and flue systems of heating is adopted, by which economy of fuel to a considerable [Pg 56] extent is effected. The boiler is so set that the back of it and all the connecting pipes are inside of the house, only the fire and ash pit doors project through the brick partition into the boiler pit. Much heat is generally wasted from hot water boilers by the direct connection of the chimney with the outer air, that might be saved by means of a well constructed flue. It will be seen that the smoke from the boiler is carried under the tank, in this instance through 8 inch vitrified drain pipe. To prevent the cracking of the pipe near the boiler the first 6 or 8 feet is laid with cast iron pipe. Wooden tanks built on posts and elevated two feet above the floor furnish bottom heat. These tanks are two feet six inches wide and six inches deep, built of 1-1/4 inch pine, well put together with white lead and securely nailed and screwed. A division through the centre separates the flow and return water. Roofing slate of proper size is used to cover the top, the joints of which are carefully cemented to prevent the escape of steam. Sand is placed directly on the slate as a plunging material for the pots containing cuttings. In the south division tanks are also used, but as the plants are potted off when placed there, bottom heat is not so necessary; the sand is dispensed with and the pots rest on a shelf or table built about two inches above the tanks, allowing the heat radiated from the slate to diffuse itself through the [Pg 57] house. Slides in each tank afford means of shutting off the water allowing each house to be worked independently. The centre of house is occupied by an earth bed in which the plants (after becoming well rooted in the small pots, to which they are first transferred from the cutting pots) are carefully transplanted and will form large and vigorous vines by the end of the season.

Fig. 5.—*Section of Propagating House.*

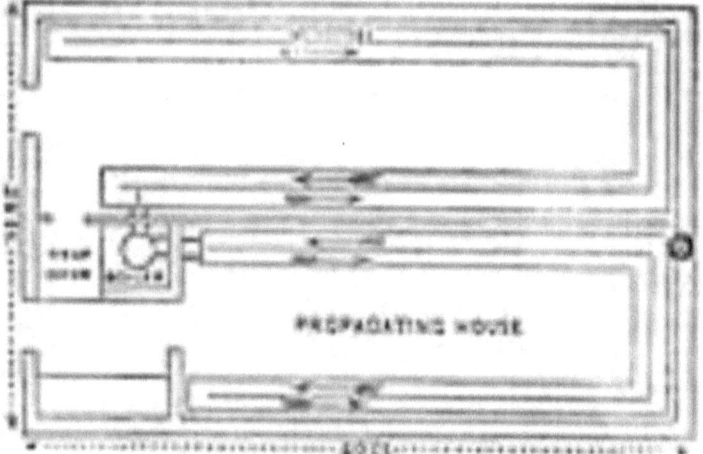

Fig. 6.—*Plan.*

DESIGN No. 2.

In Figure 7 is given a perspective view of a propagating house of an ornamental character. It is intended for forcing early vegetables, strawberries, grapes in pots, and such general propagation of plants as are needed on a country place of moderate extent. The curvilinear roof gives beauty to the design as well as affording more head room inside than the ordinary straight rafter.

Fig. 7.—*Perspective View.*

The pitch of the roof is quite flat, a straight line between the ends of the rafter forming an angle of only 28 degrees with the horizon. It was desirable to have the roof as low as was consistent with sufficient head room, that the plants might be as near the glass as possible, without the necessity of high stag [Pg 58] ing in the centre. The house has the ends to the east and west. At the west end is an ante-room, not shown in perspective view, containing the boiler, seed drawers, desk, &c. On the north side of house are beds for propagating plants, and the south side is used for early vegetables, strawberries, &c. In the centre is a large bed of earth used for grapes in pots, vegetables and plants. A portion of the roof on the south side can be raised when it is desirable to [Pg 59] harden off the plants in spring. The foundation is of wood, locust posts being used, with boards nailed upon both sides and coated with coal tar. The house is forty one feet long and sixteen feet wide, and is heated by a tank constructed as follows: brick piers are built three feet apart on which are laid common blue flag stones six feet long and two feet wide. The sides and [Pg 60] divisions of the tanks are built of brick, and cemented inside. One of Hitchings & Co.'s boilers furnishes the heat, and is connected with the tank by two inch iron pipe. Above the tanks are the propagating beds as shown in figure 8. The tank, with the exception of that part across the end of the house is covered with beds and no provision is made for other heat than that radiated from the sides, and that portion left uncovered at the end. In the practical working of the house, this has been found insufficient, and pipes have been introduced for atmospheric heat, the tanks being still retained for bottom heat. [Pg 61]

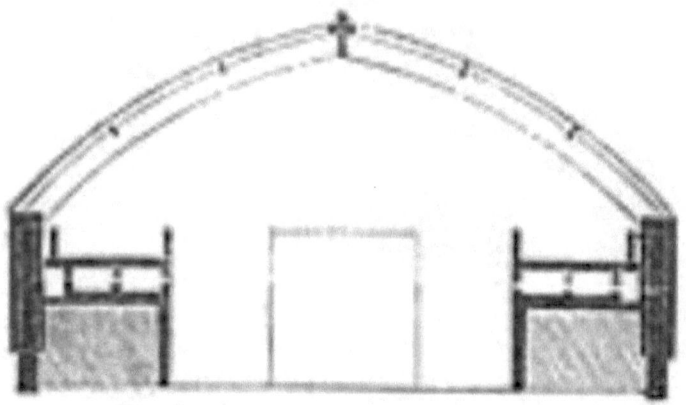

Fig. 8.—*Section.*

Fig. 9.—*Ground Plan.*

Fig. 10.—*Perspective View.*

DESIGN No. 3.

[Pg 62]
[Pg 63]

The following plan is similar to the one previously given, and was erected for the same general purposes. It has, however, been found to answer so well for a general green-house, that there is but little forcing or propagation carried on. At the east end is the boiler pit, seed room, &c.; the roof of which is of tongued and grooved boards bent to the curve of the roof and battened. The foundation is of stone, and the whole house of a substantial character. Bottom heat is furnished by brick tanks built in the same manner as before described, the water in which is heated by iron pipes running through the tanks (see section *Fig.* 12.) The pipes being also used to heat a grapery near by on a higher level, it was necessary to carry them thus. This arrangement for bottom heat is not as good as when the water flows directly into the tank from the boiler. There is a large bed in centre of house in which pots of plants are plunged, and considerable shelving at ends of house. Bottom ventilation is obtained by six inch earthen drain pipe, placed on a [Pg 64] level with the floor inside and running through the wall and up to the surface of the ground outside, where they are covered with wooden caps for regulating the amount of air required. Ventilators are placed over the doors and in the opposite end of house, in addition to which, the sash in the doors are hinged and can be opened when needful.

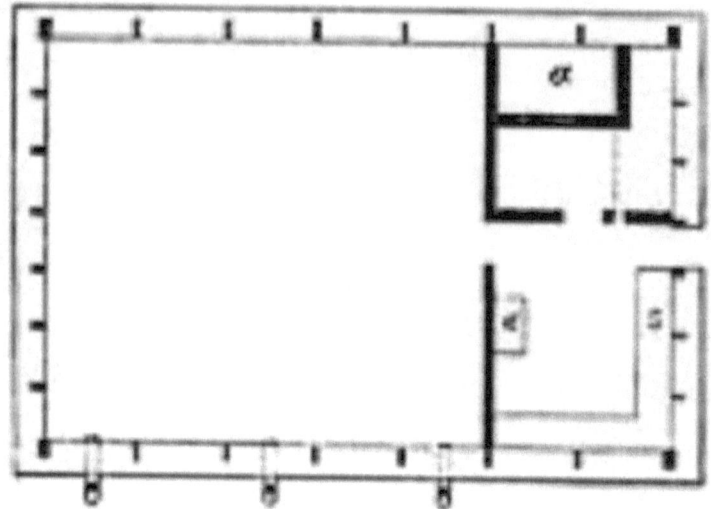

Fig. 11.—*Ground Plan.*

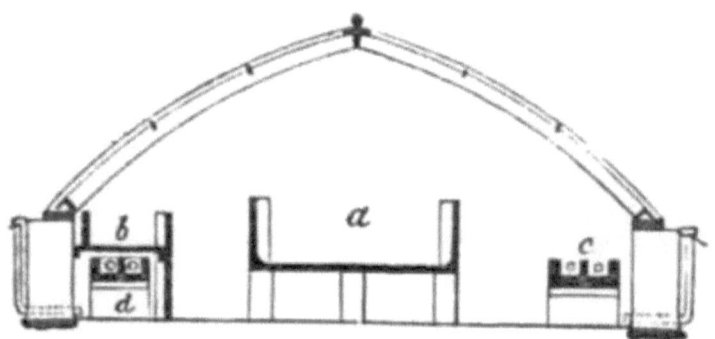

Fig. 12.—*Section.*

DESIGN No. 4.

This design combines a grapery, and forcing, and propagating house in one. *Figs.* 13, 14, 15, show side elevation, south front, and section through the centre. The dimensions are twenty feet in width by forty three feet in length, to which ten feet have since been added, enclosing boiler pit C. and potting room not shown in sketch. The foundation is built on locust posts with plank nailed upon both sides. Such foundations we do not advocate, as they are a bill of expense, for needful repairs, every four or five years, and the additional outlay for permanent brick or stone foundations is money well invested. In the present case, the owners wishes were carried out. On the ground plan, that part designated A. is devoted to the growth of grapes. The border is all inside of the house and is about three feet in depth. At the dotted line a wall is built across the house to sustain the border, the floor of B. being two feet lower. The central portion of B. is devoted to grapes in pots. At the sides of B. are beds for propagating plants, forcing vegetables, &c., furnished with bottom heat from brick tanks which extend entirely around the house and heat the grapery part as well.

Fig. 13. — *Side Elevation.*

Fig. 14. — *South Front.*

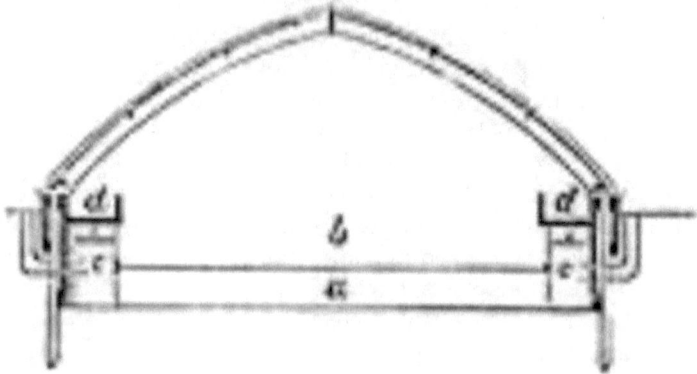

Fig. 15. — *Section.*

Pipes laid underground from the outside furnish fresh air when desired and ventilation in the roof is also provided for. [Pg 67]

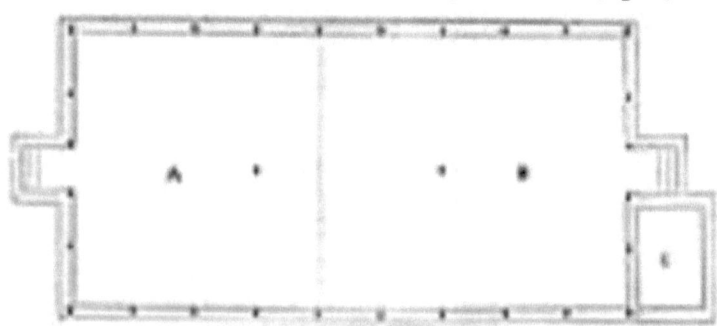

Fig. 16. — *Ground Plan.*

[Pg 68]

DESIGN No. 5.

The following design is a house with a straight roof of low pitch, and was built with considerate regard to cost, for which reason, among others, the foundations are of wood, and side lights are omitted. The sides are of inch and a half plank nailed to locust posts, the space between the inside and outside lining being filled with charcoal dust. Such foundations do very well at first, but the wood in contact with the ground will decay in three or four years, and require repairs—though locust posts will last for many years.

This house is quite narrow, being only twelve feet wide. It has tables on either side and a walk in the middle, through which is a row of light posts to support climbing plants. Ventilation is effected at the ridge by six ventilators. There are also ventilators over and in the doors. The house is heated by two four inch pipes under the tables. The boiler pit is located in a sunken shed outside, not shown in the plan. This house has been used for growing such plants as are generally found in an amateur's collection, and has given satisfactory results. [Pg 69]

Fig. 17.—*Perspective View.*

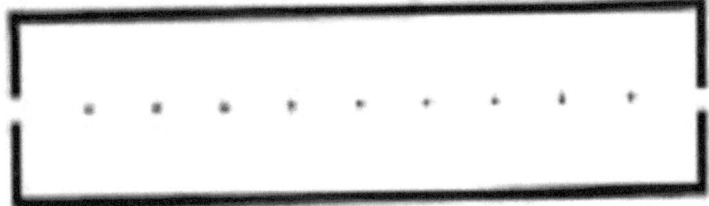

Fig. 18.—*Ground Plan.*

[Pg 70]

DESIGN No. 6.

Our next illustration is of a green-house and grapery combined, seventy feet in length by twenty feet wide. It is divided by a glass partition into two compartments, either of which can be heated at pleasure from the same boiler, by means of cut-offs in the pipes. This house was designed to be heated entirely by the tank system, but pipes were afterwards substituted except for the propagating beds. This house is located on a large village lot at Kingston, N. Y., near the dwelling, and is in full view of the street. The exposure is all that could be desired, and the protection from northerly winds perfect. A boiler pit is located outside, at the side of the building, over which a handsome summer-house is built which shields it entirely from view. The foundation is of brick, and the whole workmanship is first class. The side sashes are three feet high, and each alternate one is hung for bottom ventilation. There are also the usual ventilators in the roof. [Pg 71]

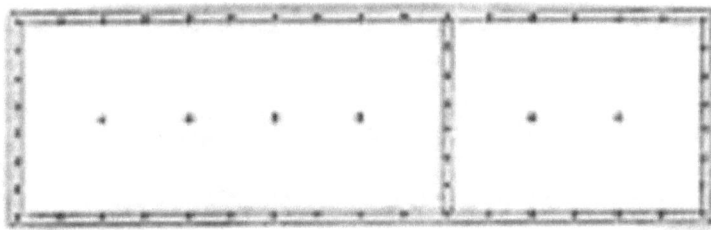

Fig. 19. — *Ground Plan.*

[Pg 72]

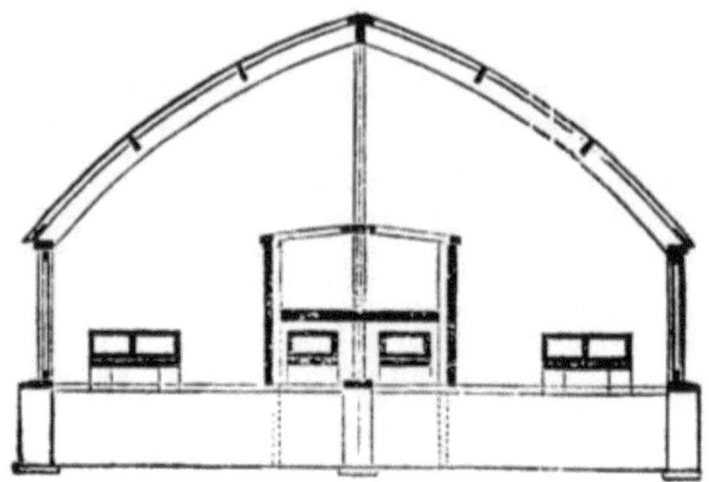

Fig. 20.—*Section*.

Fig. 21.—*South Front*.

[Pg 73]

DESIGN No. 7.

Fig. 22.—*Perspective.*

Fig. 23.—*Section.*

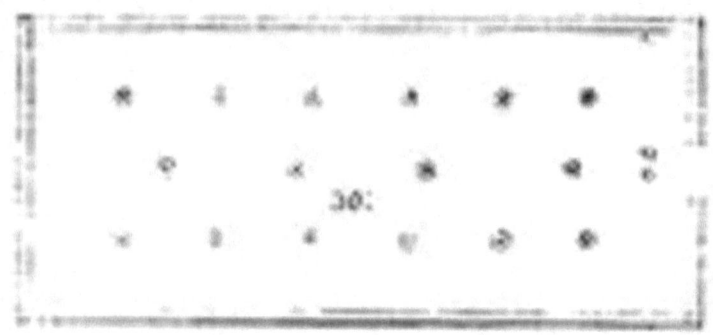

Fig. 24.—*Ground Plan.*

This design is for a Cold Grapery of low cost. The object contemplated is to secure a house that shall answer the purpose intended, and be a complete working house in all its parts, without unnecessary expense. The general outside appearance, Fig. 22, is similar to a plant house before illustrated, the straight roof affording little opportunity for architectural variety. By referring to Fig. 23, section, and Fig. 24, ground plan, it will be observed that rafters to support the roof are dispensed with, except two at each end to form the verge and finish. The [Pg 74] ridge and purlins are supported by light 2x3 inch posts, which rest upon larger posts beneath the ground. This is a considerable saving, both in material and workmanship. Posts set three feet into the ground form the foundation for the sides and ends of the house. The sides are two feet above the ground, and the entire structure is but ten feet in height, enabling the gardener to reach nearly every part of the roof from the ground. The posts may appear to be an objection, but in practice they are found not to be so, but are useful to train the vines upon. Five rows of vines are planted, two in the usual manner at the sides, and one row along each line of posts. The object in plant [Pg 75] ing thus, was to get as much fruit as possible in the shortest space of time. These centre vines will give several crops of good fruit before they will be much interfered with by those trained upon the roof. 9x15 glass was used in glazing, to lessen the expense of sash bars, the glass being laid the 15-inch way. This glass, being very true, has made a good roof, but 10x12 is as large a size as will usually be found to answer. This house is distinguished from most of our other designs by the greater amount of light [Pg 76] admitted, owing to the absence of rafters and the less than usual number of sash bars. The sides and ends are boarded perpendicularly, and battened. Ventilators are provided on each side of the ridge and over the doors, while the sashes hung in the doors furnish sufficient bottom ventilation. It was desirable to have the house raised or appear higher owing to the slight depression of the ground at the site, and for this reason the border was all made above the surface two feet and a half in height, composed largely of decayed sods, with an addition of muck, coal and wood ashes and a small quantity of stable manure. It has been found to work admirably, and preserve an even moisture throughout. Elevated borders are highly recommended by some exotic grape growers, and our experience with them is much in their favor. At present the inside border is alone completed, as it was desirable to plant the vines, and sufficient materials were not at hand to complete the whole. Vines were planted the 1st of June, 1864. [Pg 77]

DESIGN No. 8.

THE POLYPROSOPIC ROOF.

Polyprosopic is not a dictionary word, at least we cannot find it in our two-volume large quarto edition of Webster, but Loudon makes use of it to name a special form of roof sometimes made use of in the construction of Horticultural buildings, the true meaning of which we believe is, that the interior side or outline of the rafter is curvilinear and the exterior formed of planes or faces.

A very extensive practice in the design and erection of Horticultural buildings of all classes and for all purposes, from the low priced commercial shed to the finished crystal palaces, that adorn our finest country seats, has led us to a more thorough investigation of this now very important subject, and we have been enabled by a long practical experience in the construction and practical management of Horticultural buildings to reach conclusions relative to form, combination, heating and management that could not be arrived at in any other manner.

We have illustrated examples of the straight and curvilinear roofs, and now give the polyprosopic roof, in which manner we have erected some half dozen graperies and plant houses. [Pg 78]

Fig. 25.—*Perspective.*

[Pg 79]

This particular form of hot houses was described by Mr. Loudon in his encyclopedia of gardening some thirty years ago, and he says, "he considers it to be the *ne plus ultra* of improvement as far as air and light are concerned."

Mr. Leuchars in his practical treaties on hot-houses published some twelve or fifteen years since, illustrates this form of house and says: "It is by some considered superior to all other forms for winter forcing."

Fig. 26. — *Section.*

Mr. James Cranston of Birmingham, England, has also adopted this form of construction, which in many respects he considers ahead of all others. It seems to have been very generally known and used by many builders of glass-houses, and its numerous combinations of sliding, lifting, and permanently fastened sash, has been public property for upwards of thirty years. Although nearly approaching to the curvi [Pg 80] linear, form it lacks the graceful beauty of a continuous curved line, and as excessive ventilation so necessary in the climate of England, is not required in our dry sunny atmosphere, the lifting or sliding sash roof is not considered so desirable as the continuous fixed roof, which is at once the most beautiful and the most economical roof yet introduced.

The principal advantage of the Polyprosopic roof, is its portability, that is, it can be made in sashes, and transported to any portion of the country, thus obviating the necessity of painting and glazing in the hot atmosphere of the interior, or loss of time from storms, etc., on outside work. The fixed roof house can be sent anywhere primed, but the glazing and second coat of paint must be done after the erection of the building; either house we think equally well adapted to growing purposes, but as a matter of beauty and economy we give the preference to the fixed curvilinear roof.

The engraving is a view of a Plant House, erected by us for Mr. Geo. H. Brown, on his beautiful estate of Millbrook, near Washington Hollow, Duchess County, New York. The plan of the house gives two nearly equal apartments, one to be used as a propagating and forcing house, and the other as a conservatory or show house for plants and flowers. Both are heated by the circulation of hot water and can be worked [Pg 81] independently of each other. Such houses add very much to the attractions of a country estate, and impress a stranger with a higher degree of taste and refinement, while the owner has added very much to his luxuries and enjoyments.

DESIGN No. 9.

Fig. 27.—*Perspective View.*

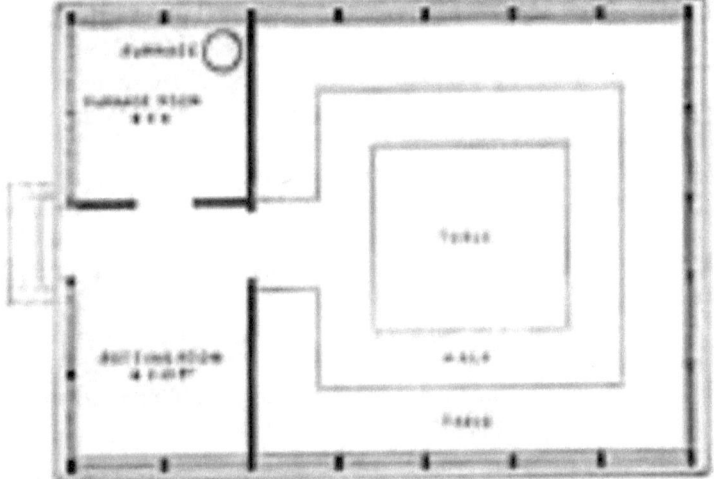

Fig. 28.—*Ground Plan.*
[Pg 82]
[Pg 83]

In this design we give a small Green House which has been erected in a substantial and permanent manner. The Green House is quite small, being only 20 by 30 feet. It is intended to keep bedding plants, Camellias, Oranges, and similar things, during the winter, and also to propagate such plants as may be wanted for bedding purposes on a place of moderate dimensions. This house runs east and west. Its position was determined partly by the nature of the ground, but mainly by the propagating bed. *Fig.* 28 is the ground plan. The large compartment is nearly twenty feet square. The potting-room, which is at the west end of the house, is eight by ten feet, and is fitted up with desks, drawers, and other necessary conveniences. The furnace pit, at the same end of the house, is eight by eight feet, and contains ample room for coal. The house is heated by two [Pg 84] four-inch pipes. The large compartment has a side table for plants. On the north side of the house there is a propagating bed, the bottom heat for which is supplied by a hot-air chamber. This hot-air chamber is formed by simply inclosing a portion of the iron pipes. In the plan there is a large table in the centre of this compartment; but this was not put in, the owner adopting the suggestion of setting his large plants on the floor of the house; a very excellent plan in itself, but which was subsequently very much marred by filling in the whole floor of the house to the depth of six inches with coarse pebbles, to the injury, we think, of the subsequent well-being of the house. The idea was, an appearance of neatness, the preservation of the tubs, and to prevent the roots from running through; but an inch of nice gravel would have secured the first without the objections that lie against the thick coat of pebbles, while the other objects will not be secured; for the tubs will rot, and the roots will not thus be prevented from running through the pots. This object must be secured by other means than pebbles. The pebbles are unpleasant to walk on, become heated, and dry off the house too rapidly, to the manifest injury of the plants. We merely mention the subject, that our readers may avoid a similar error, and save themselves the money thus needlessly spent. [Pg 85]

Fig. 72 is a perspective view of the house. The west end is boarded and battened. This corresponds with the general design of the house, and presents a neat finish. The sides, except the potting room, are of glass, the sashes being about three feet high. Every other sash is hung at the bottom, for the purpose of ventilation. The roof is a continuous glazed roof, and is quite flat, which is a decided advantage to the plants within. There are no ventilators in the roof, the top ventilation being effected by means of the sashes over the doors at each end, which are hung at the bottom for this purpose, and afford abundant ventilation for a house the length of this one. There is an ornamental crest along the ridge, and at each end a neat finial.

DESIGN No. 10.

Fig. 29.—*Perspective View.*

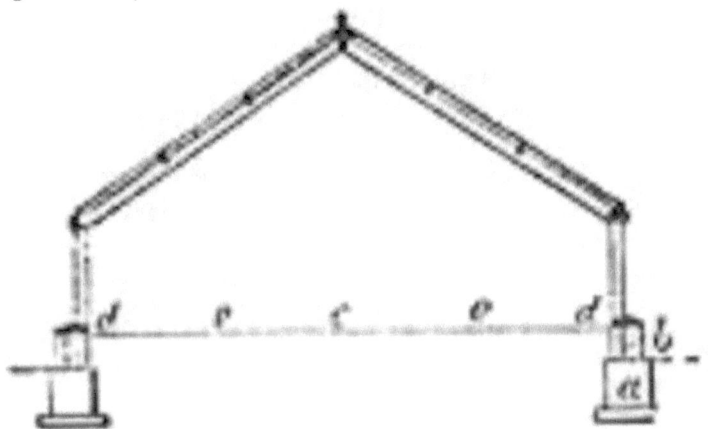

Fig. 30.—*Section.*

Our next example is a Cold Grapery, erected at South Manchester, Connecticut.

[Pg 86]

Fig. 29 is the perspective view of the house, and *Fig.* 30 is a section. The house is twenty feet wide and sixty feet long. In *Fig.* 30, *a* is a stone wall, with a drain under it. *b* is a hollow brick wall. *d, d,* is the ground level of the house on the inside; the line be [Pg 87] low *b* is the level on the outside, but the earth is embanked against the brick wall to within an inch of the sill. A small house is shown at the north end which is used for tools, potting, &c. The border is about three feet deep, and occupies the whole interior of the house. There is no outside border. On the bottom is placed about one foot of "tussocks" from a neighboring bog, which may in time decay. The border is made up pretty freely of muck, with the addition of sand, loam, [Pg 88] charcoal dust, bone dust, etc. There is a row of vines, two feet and a half apart, at each side of the house, at *d, d*. There are two other rows at *e, e*. There are also a few vines at *c*, and at the ends of the house. The rows at *d, d*, form fruiting canes half way up the rafters; those at *e, e*, go to the roof with a naked trunk, and furnish fruiting canes for the other half of the rafters. The fruiting canes are thus very short, and easily managed. The house was planted in the month of April, with such grapes as [Pg 89] Black Hamburgh, Victoria Hamburgh, Wilmot's Hamburgh, Golden Hamburgh, Muscat Hamburgh, Chasselas Fontainebleau, Frontignans, Muscat of Alexandria, Syrian, Esperione, Tokay, and some others. The plants were very small, and the wire worm injured some of them so as to make it necessary to replant; but the growth of those not injured was very good. A fine crop of Melons, Tomatoes, Strawberries, etc., was taken from the house the first year. The second year a few bunches of grapes were gathered, and every thing went on finely.

Fig. 31.—*Ground Plan.*

This is the third year in which the house has been in operation. Our last visit was in the early part of August, 1863, when we counted 734 bunches of grapes, weighing from one to seven pounds each, the Syrian being the grape which reached the last figure. Almost as many bunches were thinned out. In some cases too many are left, but they look very fine. The Muscats are extremely well set, and some of the bunches will weigh fully three pounds. The Black Hamburghs look quite as well; but the finest show of fruit is on the

Esperione. The large number of bunches is owing to the manner of planting; so many could hardly be taken the third season from a house planted in the ordinary way. The canes, it will be borne in mind, are now only fruited about half their length.

The exposure of this house is a very bleak one, and [Pg 90] the climate cold and fickle. In order to provide against a late spring frost, a coil of one inch pipe was inclosed in brick work, with a fire chamber under it. From this coil a single one inch pipe was carried around the house next the side sashes. It is found to answer the purpose, having on one occasion kept the frost out of the house, when the crop in the house of a neighbor was destroyed. In many places, some resource of this kind is necessary, and a small boiler with a single pipe will in most cases prove sufficient.

DESIGN No. 11.

The following illustration is of a Plant House attached to a dwelling, and is quite different in its plan from those before given. It was designed and erected for J. C. Johnston, Esq., of Scarborough, N. Y.

It is built on the south side of the dwelling, and is entered from the parlor as well as from the pleasure grounds. *Fig.* 32 is a perspective view, which gives the reader a good idea of its general appearance, though we can not help saying that in this case, at least, the picture does not flatter; the house looks finer on the ground than in the picture. The circular house on the southeast corner is strictly an ornamental feature, and a very pretty one. [Pg 91]

Fig. 32. — *Perspective View.*

[Pg 92]

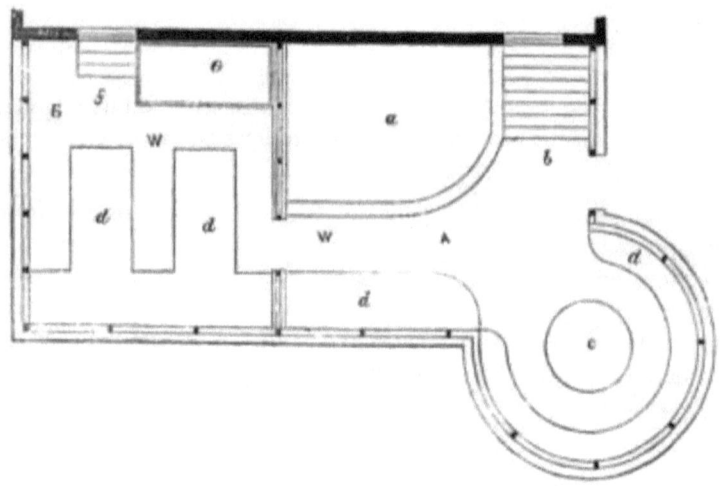

Fig. 33. — *Ground Plan.*

[Pg 93]

The interior arrangement is shown in the ground plan, *Fig. 33.* The house is divided into two compartments, A and B. The last is intended for growing and propagating plants. The house is heated by hot water pipes, the boiler being placed in the cellar of the dwelling, which is entered by the steps, *f; e* is a propagating tank, fitted with sliding sashes. It is quite large enough to propagate all the plants the owner will want; *d, d,* are beds about a foot deep, with a moderate bottom heat, for plunging pots in when desired; *w* is the walk. This compartment is to be used for bringing plants into bloom, after which they are to be taken to the show room or conservatory, marked A in the plan. The arrangement of this compartment is such, that all the plants in it may be seen from the parlor door or window, the steps leading to which are marked *b; a, d, d,* are tables; *c* would make a pretty little fountain, but it is intended at present to put it in the form of a rustic basket, and fill it with ornamental plants. The effect can not be otherwise than good. Climbing plants of various kinds will be trained up the mullions and rafters of the circular house, and allowed to hang in festoons from the roof. When the house is filled with flowering and ornamental-leaved plants, with climbers dependent from the roof, the effect will be charming. [Pg 94]

DESIGN No. 12.

COLD GRAPERIES FOR CITY LOTS.

In this illustration is given three graperies, designed and constructed by us for Mr. John H. Sherwood of this city, which are among the first, if not the first erected in New York, as an elegant, substantial and attractive addition to three very superb palatial residences on Murray Hill, near 5th Avenue. These latter are buildings, such as, in style and workmanship, very few persons in this country, outside of New York, have seen, and such as but few of the first class builders of New York are competent to erect.

Centrally located in the aristocratic portion of a city noted for its wealth, taste and influence, these Graperies will be carefully watched as an index of what the future may do in the increased demand for houses on city lots for Horticultural purposes.

A full sized lot in the city of New York is twenty-five feet wide by one hundred feet in depth. The ground attached to each dwelling in this case is equal to two full sized lots, being twenty-five feet wide and two hundred feet in depth. The dwellings front on Fortieth Street, behind which are the yards, twenty by twenty-five feet; the Graperies, which are twenty-five feet by forty feet; then the coach houses, which [Pg 95] front on, and are entered from, Thirty-ninth Street, thus using the whole space.

[Pg 96]

Fig. 34.—*Perspective.*

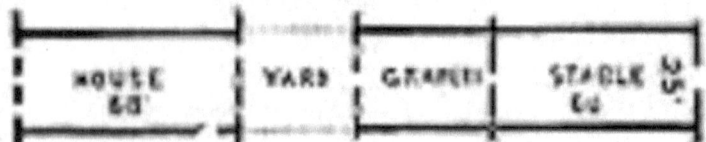

Fig. 35.—*Ground Plan.*

The graperies are intended to be used without heat; but whenever desirable, heating apparatus can be easily introduced, and the grape season materially lengthened. For practical purposes only, and on open grounds, it would, perhaps, have been better to have built the houses lower; but as grapes are usually fruited next to the glass, the principal objection to high houses for grape culture is the extra labor in getting up to the vines for pruning and training. These houses are purposely built higher than is now usual, to give a finer effect from the drawing-room windows, and to secure, as far as possible, the influence of the sun's rays.

By the use of glass houses on city lots, much enjoyment may be had by all who have a desire to spend their time in growing fine fruits and flowers. Pot vines and trees condense a vineyard and orchard into a wonderfully small space, and border vines yield a harvest of glorious fruit that surprise all not accus [Pg 97] tomed to seeing and eating such luxuries. Our city lots, with rare exceptions, are well adapted to the growth, under glass, of grapes and orchard fruit, and the forcing of vegetables. There are many of them somewhat shaded during portions of the day, yet the better protection is something of a compensation, and besides that, it is still an open question whether sun-light is alone essential in perfecting fruit; daylight in many cases does pretty well.

The failure to receive the sun's rays the entire day would not deter us one moment from the erection of a horticultural building. Those who grow fruit where all conditions are most favorable to success, do not enjoy the same pleasure nor attain the same skill as those who battle with difficulties; success easily acquired has not the same value as success which is reached by persistent effort against adverse circumstances.

Unlike the garden of a country gentleman that blossoms and fruits and passes away in a season, the horticultural building properly heated is a perpetual pleasure, a garden the year round; vegetables and fruit and flowers follow each other without intermission.

Very much is due to the foresight and energy of Mr. Sherwood, in inaugurating the introduction of horticultural structures of this class in New York. [Pg 98] Few gentlemen of wealth have had the same opportunity, and few less would have the courage to take the first bold step in this matter. It cannot, however, by horticulturists, be looked upon as an experiment, however much those inexperienced in such matters may be disposed to criticise.

We are sure that Mr. Sherwood has done something that will advance the cause of Horticulture, and equally sure that he will be successful in the result. We shall feel much interested in his progress.

DESIGN No. 13.

Fig. 36.—*Perspective.*

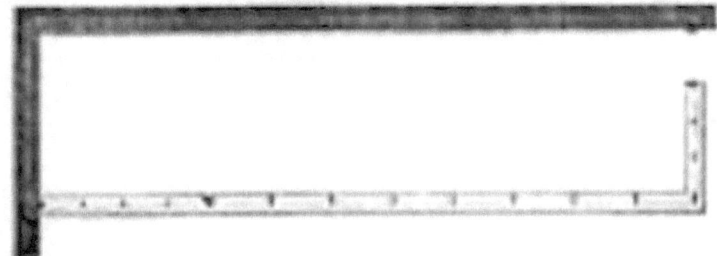

Fig. 37.—*Ground Plan.*
[Pg 99]
[Pg 100]

In our present illustration we have an example of what may be done with a wall. It was necessary, for certain purposes, to cut away an embankment, and build a sustaining wall. After this had been done, we were asked if the wall could not be devoted to some useful purpose, and it was determined to build a lean-to grapery against it. The chief difficulty in the way was the wet and springy nature of the ground at the level marked water line in *Fig.* 38. It was found, however, that it could be drained; but at certain seasons of the year surface water would accumulate from the overflow of a milldam. But there is gener [Pg 101] ally some way to overcome difficulties. In this case, the border was placed inside the house, and well raised, with a firm concrete bottom between the ground and water lines, and suitable drains connecting with the main drain under the front wall, to secure the requisite degree of dryness inside. Up to the present time we believe every thing has gone on very favorably. We have no doubt that many other places, now deemed useless, might be converted into good graperies at an expense that the results would fully warrant. In case this was successful, it was the owner's purpose to extend the house along the wall at the left; and it was therefore deemed best to insert the valley at the angle, to save future expense in tearing down the end of the house.

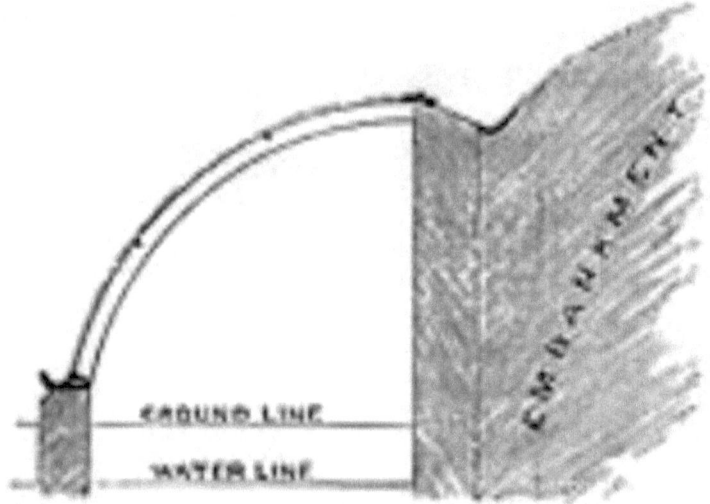

Fig. 38.—*Section*.

Fig. 36 is a perspective view of the house, which, in connection with *Fig.* 38 will give the reader a good idea of the general arrangement. *Fig.* 37 is a plan. [Pg 102]

DESIGN No. 14.

Our next illustration is a hot grapery. It is forty-one feet long and twenty feet wide. *Fig.* 39 is a perspective view. It is covered with a low, continuous, curvilinear roof, and is without side lights. The omission of side lights materially lessens the cost of the house, and secures additional warmth. In some cases, side lights serve no other purpose than architectural effect. Graperies, propagating houses, and plant houses generally may very well be constructed without them; some of these houses, indeed, are very much better without them.

In the present instance, to prevent what is called a "squatty" appearance, and also to give additional headway, the side walls were carried up some twenty inches above the ground line. The house is thus made to assume a handsome appearance. Air is introduced into the house at the sides, through underground wooden air chambers opening on the inside near the walk. Instead of these wooden air chambers, we now use six inch glazed pipes, as being more convenient and durable. It is an effective and excellent mode of introducing fresh air, without letting it directly on the plants. Ventilation is effected by the sash over the end doors, and also by ventilators placed along the ridge board. [Pg 103]

Fig. 39.—*Perspective.*

[Pg 104]

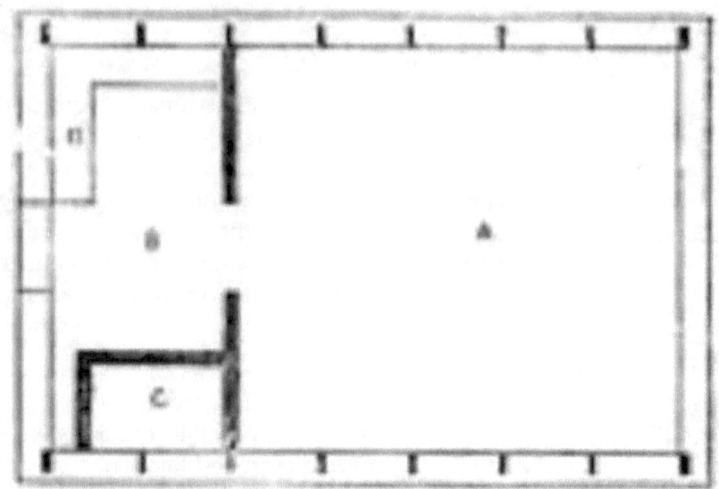

Fig. 40.—*Ground Plan.*

Fig. 40 is the ground plan. At the north end a small room is partitioned off for a boiler pit. On one side is a chest of drawers for seeds, &c., and on the other some shelving. In connection with the boiler pit is a coal bin, not, however, of very large capacity. The house is heated by two four-inch pipes, the design being not to work the house very early. The border is entirely inside the house, and is composed principally of sod, muck, and gravel, with the addition of some old manure and bone shavings. The vines have done well, annually ripening a fine crop of fruit, and the house has in all respects proved to be satisfactory. [Pg 105]

DESIGN No. 15.

This is a plan of a range of houses designed and built for Joseph Howland, Esq., of Matteawan, N. Y. It is a large and imposing structure, befitting the character of Mr. Howland's ample grounds. It stands at the north end of the kitchen garden, and conceals it from the dwelling, from which the range is in full view. A part of the structure on the right, used as a green house, not shown in ground plan, was built some four or five years ago with the old sliding sash roof, which was found so unsatisfactory that at the time of the erection of the new portion, this roof was removed and replaced with a curvilinear fixed roof to correspond with the rest.

It will be observed that the range is divided into two parts by a road-way. The design of this was to enable the family to visit the houses at any time in the carriage without exposure to the weather, and enjoy the fruits, flowers, and temperature of tropical climates, without the necessity of leaving their homes.

The north side of the middle houses is covered with boards and battened. End ventilation being impracticable here, top ventilation is increased so as to meet all requirements. [Pg 106]

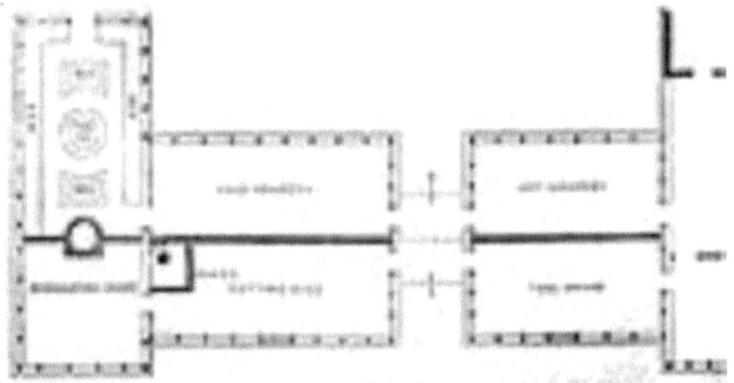

Fig. 41.—*Ground Plan.*

[Pg 107]

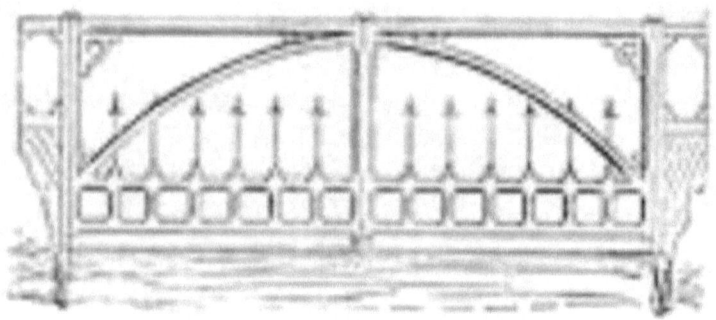

Fig. 42.—*Double Gate*.

[Pg 108]

Fig. 41 is the ground plan. On the right is the old green house, a portion of the foundation of which is shown. This communicates with the hot grapery and tool house, under which is a capacious root cellar. From the covered road-way, all the parts of this extensive range are easily accessible. Across the road-way, between the houses, is a handsome double gate, a sketch of which is given in *Fig.* 42.

Crossing the road-way, we enter the cold grapery. The foundation of this rests on piers, the border being outside. There are hot-water pipes in this compartment, to be used only to keep out frost. It may, however, be used as a "second" hot grapery. Passing out of the cold grapery, we enter what may be called the conservatory, its principal use being for the show of ornamental plants; and to this end it has several accessories which add much to its beauty. One of these which may be noticed is a neat fountain in the centre; always a pretty feature wherever it can be introduced. Another is a rustic niche or alcove in the north wall, built of rough stones, over and through which the water constantly trickles into a basin. Its full beauty will not be seen till it has acquired age, and become covered with mosses and ferns. Fortunately for the plants and for good taste, there is no shelving in this house. Beds are formed of brick, with a neat coping, in which the pots are set. [Pg 109] This arrangement is much more effective than any manner of staging could possibly be.

Fig. 43.—*Interior View.*

[Pg 110]

In order to give the reader an idea of the interior of this apartment, we have prepared a perspective view of it. (See *Fig.* 43.) From this a good conception can be formed of the appearance and arrangement of the beds, fountain, &c.

Returning through the cold grapery, we have on its north side a boiler and potting room. The boiler pit is sunk beneath the floor of this room, and has connected with it a coal bin and shoot. Communicating with the potting-room is a propagating room, in the north end of the conservatory, and divided from it by a solid partition. It is provided with hot-water pipes for furnishing bottom heat. It will propagate all the bedding and other plants needed on the place. It will thus be seen that there are ample facilities for furnishing an abundant supply of grapes and flowers. The house, as a whole, forms a marked feature of the grounds. [Pg 111]

DESIGN No. 16.

The following design was prepared for Dr. Butler, of the Retreat for the Insane at Hartford, Conn. The doctor had conceived the idea that a green-house might be made to serve a very important part in the treatment of the insane, having noticed the soothing influence of plants upon his patients, more especially the females. We have no doubt that his anticipations will be fully realized; for we can scarcely conceive of anything better calculated to heal the "mind diseased," than daily intercourse with these voiceless, but gladsome children of Nature.

Fig. 44 is a perspective view of the house. It is twenty-four feet wide and seventy-five long. It has a low, curved roof, and side sashes three feet six inches high. We do not make these roofs low for the sake of architectural effect, though this point is certainly gained; but rather for the sake of the plants, a low roof, in this respect, possessing incalculable advantages over one that is steep. When attention is once generally fixed on this point, plant growers will not be slow to acknowledge the superiority of the low roof. It has often surprised us that gardeners will assume a great deal of unnecessary labor for the sake of an old prejudice. Some of them are slow to avail [Pg 112] themselves of improvements that not only lessen their toils, but bring greater certainty and pleasure to the pursuit of their profession. Others, again, are quick enough to avail themselves of every facility brought within their reach. We could wish that the latter class might multiply rapidly.

[Pg 113]
[Pg 114]

Fig. 44. — *Perspective View.*

Fig. 45. — *Ground Plan.*

One of the prettiest features about this house is its rounded ends. The pitch of the roof and the width of the house are such, taken in connection with the circular ends, that all the lines flow into each other with the utmost harmony. These different parts were studied with reference to producing this result, and we think it has been done with some degree of success. The finials, the ornament along the ridge, and the entrance door, are all in keeping with the rest of the structure.

Fig. 45 is the ground plan. This presents some peculiarities. The house being designed for the use of the insane, it was desirable to place the heating apparatus out of their reach; the boiler is therefore placed under ground. For this purpose a vault of sufficient size to hold the boiler and several tons of coal, is built under ground in front of the house. It is substantially built of brick, and arched over. The smoke shaft is carried up through the roof, and finished above ground in the form of a column or pedestal, surmounted with a vase, as seen in *Fig.* 44. [Pg 115] To harmonize the grounds, and conceal the purpose of this column, another is placed on the opposite side of the path. In summer, these vases will be filled with plants, and the columns are intended to be covered with vines, thus making them subserve an ornamental purpose. There are two entrances to the boiler vault, one from within by a concealed trapdoor, and the other from without. The house will be heated by hot water pipes.

There will be neither shelves nor tables in the house. The plants will be set either on or in the ground, and the whole interior made to resemble as much as possible a flower garden. The plants will thus be easier seen, better enjoyed, and more appreciated than if placed either on tables or staging. In any well-designed house, the plants look and grow infinitely better upon flat tables; and a large class of plants will grow even better upon the earthen floor of the house.

DESIGN No. 17.

Our next example is a lean-to grapery for early forcing. It was designed for a gentleman in Connecticut, and we believe has since been built. [Pg 116]

Fig. 46 — *Perspective View.*

[Pg 117]

Fig. 47. — *Ground Plan.*

[Pg 118]

Fig. 46 is a perspective view. It runs east and west, and is designed to correspond in a measure with another house on the place, though the roof of this is much flatter. There are no side lights. Ventilation is effected by openings along the ridge, and by the sashes over the doors, which are hung for the purpose. The roof is continuous, and both ends of the house are glazed.

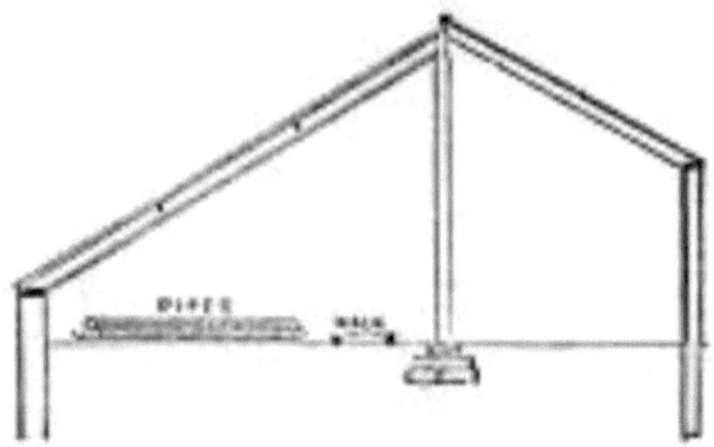

Fig. 48.—*Section*.

Fig. 47 is the ground plan. The sills of the front or glass part rest on brick piers, to allow the roots of the vines to run out, the border being both in and outside the house. A wooden partition on the north side of the walk divides the house into two unequal parts, the north being used for a potting shed, tool house, etc. This apartment is furnished with tables, etc., and is well lighted by windows at the side and [Pg 119] ends. A water tank is conveniently placed in the middle. In the northwest corner is the boiler pit. This is sufficiently large to hold coal, and is furnished with a shoot for throwing it down. The grapery is to be heated by four rows of pipes, the object being to force early.

Fig. 48 is a section, showing the arrangement of pipes, walk, etc.

DESIGN No. 18.

Plant houses having a specific object in view, it is not possible to indulge in a great variety of forms without sacrificing their utility, or creating a great deal of room that can not be applied to any useful purpose whatever. In this respect they differ in a marked manner from dwelling-houses, which allow of great latitude in design and construction. That some degree of picturesqueness, however, is consistent with utility, we think will be apparent on examining the design herewith presented. The plan was made for H. B. Hurlbut, Esq., of Cleveland, Ohio. It is intended for a green-house and hot-house combined. It is located near the dwelling and in sight of the public highway. It is in the form of a cross. [Pg 120]

Fig. 49.—*Perspective View.*

[Pg 121]

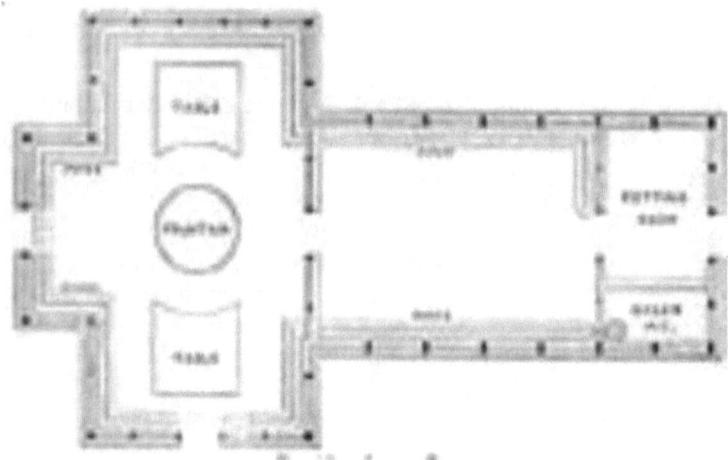

Fig. 50.—*Ground Plan.*

[Pg 122]

Fig. 49 is a perspective view, as seen from the street. The porch or front entrance is ornamented, but with an entire absence of heavy wood work. The finials and crest along the ridge are light, and harmonize with the general design. The valleys and angles break up the structure in a very pleasing and effective manner, and the elevation, as a whole, is one that will arrest attention.

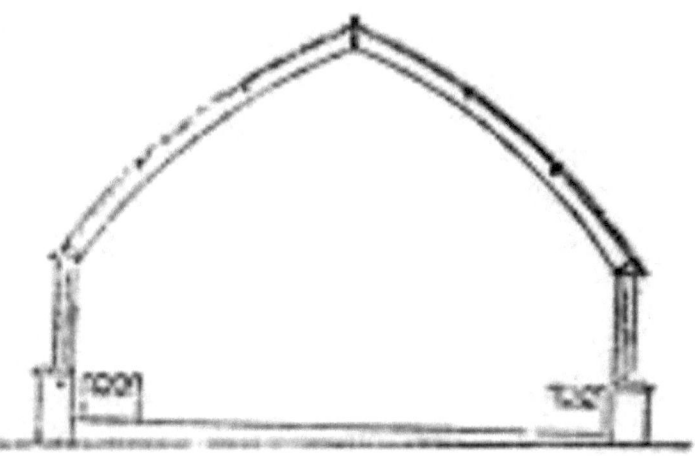

Fig. 51.—*Section*.

Fig. 50 is the ground plan. Directly opposite the front entrance is a fountain. There are two centre tables for plants, also others around the sides of the house, not shown in the plan. This apartment will be used principally for plants in bloom. The other apartment which will be kept at a higher temperature, for the purpose of forcing plants into flower. At the end, on the right-hand side, is the boiler-pit, which is partitioned off. It is large enough to hold two or three tons of coal. There is a coal-shoot on the out [Pg 123] side. On the left is the potting-room. This will be fitted up with a writing desk, and shelves and drawers for books, seeds, etc. Every other side-sash is hung at the bottom for ventilation. There are also ventilators on the top, and over the doors. *Fig.* 51 is a sectional view of the house.

There is scarcely any part of this structure that does not, at some time during the day, receive a portion of the sun's rays; some more, some less. A little judgment, therefore, on the part of the gardener who has charge of the place, will enable him to grow well a large variety of plants.

DESIGN No. 19.

Fig. 52.—*Perspective.*

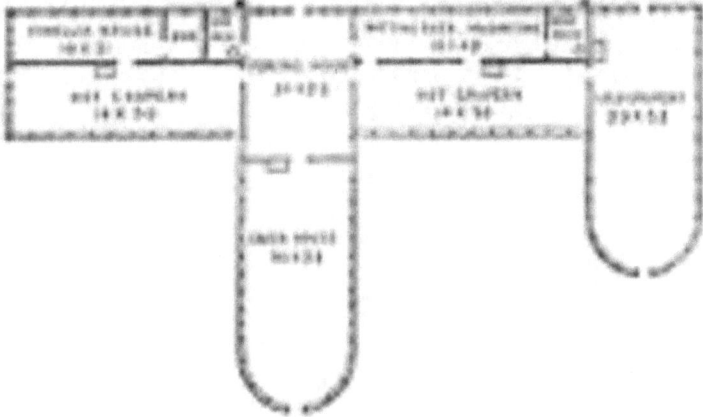

Fig. 53.—*Ground Plan.*

This design is of a plant-house of larger dimensions than any we have heretofore given. Its form was determined by its location. *Fig. 52* is a perspective.

The principal building runs east and west. This is divided by a brick wall into two unequal parts, that facing the south being the largest. On the north side we have first, at the west end, a small

Camellia house. It would be also adapted to Orchids, Caladiums, Begonias, Ferns, and all plants requiring partial shade. Next we have a moderate [Pg 126] sized bed-room for the man who attends to the boilers, one of which is in the next room. These two rooms are covered with boards bent to the curve of the roof and battened instead of glass. On the south of these three rooms is a hot grapery, to be used as a "second" house. Next, on the east, is a house designated "Forcing House" in the plan. (See *Fig.* 53.) It should be "Hot House," as this room is not adapted to forcing purposes. It is intended for plants that require a high temperature to keep them in good health. East of this is a room, or a "potting shed." Being covered with glass, it is well adapted to growing Mushrooms, propagating plants, &c., all the room not being needed for potting purposes. By the side of this room is another boiler room, and on the south another Hot Grapery, to be used as a "first" house. Then, on the east is the Cold Grapery, of goodly dimensions. Last of all we have a Green-house of large size south of the Hot-house. Thus, under one roof, we have all that is needed on a large place. We do not wish to be understood as saying that it is always best to put these houses in this particular shape; but where money is no particular object, and architectural effect is sought for, this form gives an opportunity in its broken outlines for considerable display. [Pg 127]

DESIGN No. 20.

Green-houses and Graperies are usually erected as separate structures. While it is desirable that they should be so on extensive places where much accommodation is required, in grounds of moderate extent many advantages are gained by having the houses connected. Facility for heating and management, protection of those houses requiring the most heat, by those kept cold or at only moderate temperature, and the ease with which all departments may be visited by the owner, are all obtained by such an arrangement. In the present instance the Green-house occupies a position east and west, and is protected on its north and most exposed quarter by the Grapery. The boiler located as shown on the plan, supplies heat to all the houses. The Grapery, not being intended as a forcing or early house, has but one hot water pipe, which will afford sufficient heat to enable the vines to be started two or three weeks earlier in the spring, or if not desirable to anticipate their natural growth, will prevent them receiving sudden checks from frosty nights, which sometimes happen at the latter end of April and beginning of May, after the vines have broken their buds. We can prolong the season also, until about Christmas, in [Pg 128] favorable years. Several of the late ripening, and late keeping varieties of the Grape, are intended to be grown. Lady Downes, Barbarossa, Frogmore St., Peters and others. These by the addition of another pipe and proper care in management, could be kept on the vines in fine condition until February, and perhaps March.

Fig. 54.—*Perspective.*

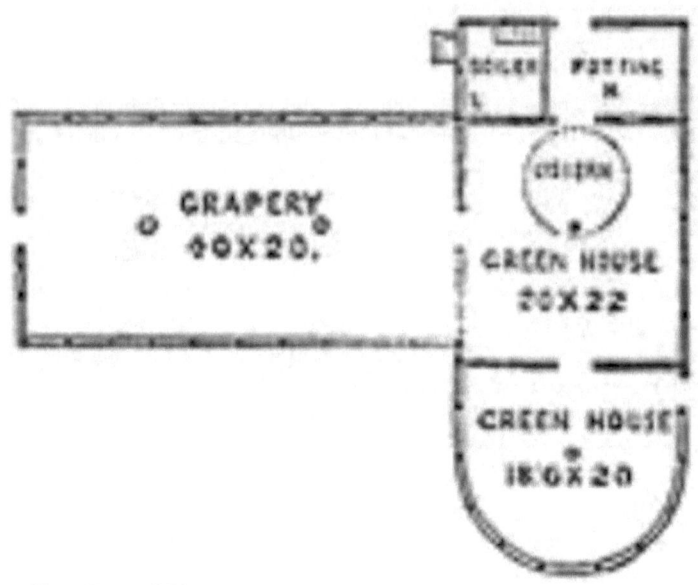

Fig. 55.—*Ground Plan.*

The sill or wall plate of the Grapery, is but two feet above the border; thus giving nearly the whole length of cane for fruiting upon the rafter. Side lights are dispensed with bottom ventilation being afforded by apertures through the brick wall, closed by shutters. The wall is supported on stone lintels, resting on brick piers placed about five feet apart, extending to the bottom of the border, allowing free access for the vine roots to the outside. Ventilation at the top is effected by means of sashes, hung in the roof at the ridge, which are raised and lowered by an iron shaft running the length of the building, with [Pg 129] elbow attachments at each ventilator. A cord and lever at one end, works the shaft, raising the whole of the ventilators at one operation. This is by far the best method of ventilation, but more expensive than that generally used. It is strong, effective, rarely requires repair, and the sashes are never in danger of being blown open and broken by high winds. The floor level of the Green-house is two feet below that of the Grapery, in order that there may be sufficient height at the sides, to place plants on the tables, and bring them near the glass. General collections of plants cannot well be grown in one house; for this reason, we have the house divided by a glass partition. By an arrangement of valves in the hot water pipes, and independent ventilation, a different temperature can be maintained in each. Plants requiring a considerable degree of heat will find a congenial location [Pg 130] in the central house, while those in bloom, and others to which a cooler atmosphere is more suitable, will be placed at the circular end of the building.

Three rows of heating pipe run around the Green-houses, which will give ample heat in the coldest weather. A propagating table is provided by enclosing a portion of the pipes in the central house. Beneath the floor is a cistern of 3,000 gallons capacity, from which tanks holding 100 gallons each are supplied by pumps. The Greenhouses are entered through a door and porch on the south, not shown in the engraving, also through potting room and Grapery. The design of these houses gives an opportunity for further addition if desired, by a wing on the south, corresponding with the Grapery on the north. Such an extension would improve the architectural appearance of the whole. An early Grapery might be thus located and be heated from the same boiler. These houses, lately designed and erected by us for John L. Rogers, Esq., of Newburgh, N. Y., form a picturesque and attractive feature in his well kept grounds, and will no doubt be a source of much enjoyment to their owner. [Pg 131]

ORCHARD HOUSES.

Glass-houses devoted exclusively to the cultivation of such fruits as are usually found in our orchards and gardens, would seem to be hardly necessary erections in our climate, with its bright and genial sunshine. But we must call to mind the almost total failure of the peach crop for several years past, on account of the severity of the winter frost, in sections of the country where this fruit was formerly cultivated with the greatest success, and ripened in the fullest abundance and perfection. We cannot forget, also, that it is next to impossible to prevent the attacks of the curculio upon our smooth-skinned fruits,—the Nectarine, Apricot and Plum—and the vast amount of vigilance and care required to counteract the invasions of the various other insect pests which visit us, and to obtain even a moderate crop, in many localities, out of doors. And we must be willing to concede that the certain means of securing even a limited supply of these delicious fruits, is worthy of our careful consideration.

Well managed Orchard houses will give us, without doubt or failure, the Peach, the Apricot, the Nectarine, the Plum, the Fig, and many other fruits in great perfection. With the addition of fire heat these [Pg 132] may be forced, and the fruit obtained much in advance of its natural season.

In England, houses for the growth of these fruits, which will generally not ripen in the open air of that climate, have been in successful use for a number of years. In these houses the trees are planted in prepared borders, which gives the roots liberty to ramble at will. The fruit thus produced is very beautiful in appearance, and if abundant ventilation is supplied, at the proper season, it is of tolerable flavor. The great difficulty in this mode of culture, seems to be in not being able to furnish adequate ventilation to the house at the period of ripening, to enable the fruit to acquire its full flavor and perfection of delicacy and richness. Another difficulty is the over vigorous growth of the trees, and the care required to restrain them within proper bounds.

An impetus was given to the erection of Orchard houses in England, by Mr. Rivers, the celebrated nurseryman and fruit grower, by the publication of his little work on the subject of Orchard houses, in which he advocated the growth of trees in pots. By this system of pot culture, we are enabled to remove the trees when the fruit begins to color, and thus to ripen and perfect it in the open air. The over-growth of wood is also restrained in this system of culture, the trees being easily managed and controlled. Great [Pg 133] success has, in many instances, attended this mode of culture in this country. Although it is but a few years since experiments were commenced here, some of our fruit growers have acquired such skill and experience, as to enable them to realize considerable profits from their investments in a money point of view, besides demonstrating the practicability of the system.

The majority of houses erected for this purpose among us, have been of the cheapest possible description. While the culture was merely experimental, this was all well enough; but now that the Orchard house has taken its place among other Horticultural structures, the same arguments we have urged against cheap Graperies will apply with equal force to this class of buildings.

The principal differences between the plans for Orchard houses and Graperies are, first, the somewhat lower roof of the former, that the pots containing the trees may stand upon the earth floor or border, while the foliage may be brought as near as possible to the glass; and secondly, the very ample ventilation required by the trees, at certain periods of their growth, and in completing the "hardening off" process of the wood, and leaves if the trees are to be removed to the open air.

Fruit trees are frequently grown in pots in Graperies. [Pg 134] After the vines have expanded their leaves maturely, and obstructed the light, it becomes necessary to remove the trees to the open air. The leaves and new grown wood being very tender, the abrupt change to a different climate is too great, and they suffer in consequence. In a well constructed Orchard house, the means of ventilation should be so ample that the trees may be gradually inured to the change; or if it is desirable to let the trees remain within the house through the summer, the access of the air must be so abundant as to give as nearly as possible that flavor to the fruit which it would acquire if fully exposed.

Fig. 56 is a perspective view of a "lean-to" Orchard house, erected some years since by J. S. Lovering, Esq. of which the following description has been furnished to us:

[Pg 135]

"Mr. Lovering's Orchard House is 165 feet long by 14 wide, is a lean-to, points south, under shelter of a hill. Back wall 12 feet high, 8 feet stone work; on top of wall 4 feet of wood, in which the back row of ventilators (2 feet by 20 inches) work, hung on rollers, and all opened and closed simultaneously by means of a wire representing a front door pull. Front wall 4 feet high, made by nailing plowed and grooved planks to locust posts, in which are cut the front ventilators, 4 feet 8 inches long by 18 inches deep, and covered by a screen of gauze wire with board shutters to close [Pg 136] tight. The roof is made of 16 feet rafters, on which rests 8 foot sash, immovable; the glass is first quality, 8 by 10. A single row of supporters on one side of the wall completes the roof. The interior is divided into three borders: the front border (3 feet 6 inches wide) is raised 9 inches above the walk (which is 2 feet 6 inches wide); the first back border is 3 feet wide, and raised 16 inches above the walk; the second back border is raised 1 foot above the front one, and is 4 feet wide. On this further back border are placed the largest trees only, having the most head room—the smallest pots standing on the front. The appearance of the house, when seen by the writer on the 7th of April, 1860, was truly magnificent, being one dense mass of bloom, (except some of the early kinds, on which the fruit was already set,) resembling a green-house of Azaleas in full flower. Peaches, apricots, nectarines, plums and figs are embraced in the assortment, and are grown principally in 11-inch pots placed about three feet apart, every leaf being fully exposed to the sun-light—vines being, of course, entirely prohibited.

Fig. 56.—*Perspective View.*

Of the success of this mode of culture in America, no one who has witnessed Mr. Lovering's house can have the shadow of a doubt.

With him it is no new experiment, having fruited pot trees in his cold graperies for several years." [Pg 137]

Fig. 57 is a section of a "lean-to" form of house, showing arrangement of trees and sunken walk to give sufficient head room.

Fig. 58 represents perspective view of a span-roofed house, in which ventilation is effected at the bottom and very freely at the ends. No ventilators are placed in the roof as they were not in this case deemed necessary.

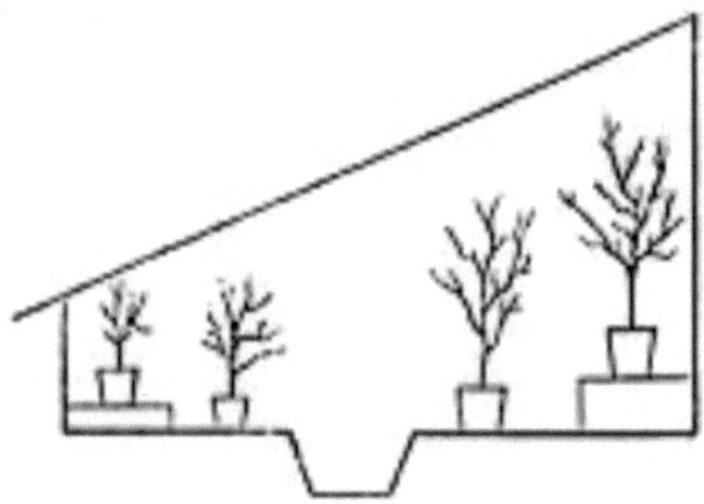

Fig. 57. — *Section.*

Fig. 59 gives a view of the interior of the span-roofed house, in which are shown the pots containing the trees. The span-roofed house we consider better adapted to the growth of Orchard fruit than the "lean-to" form, except where it is desired to force the fruit in advance of its season, in which case the lean-to possesses the advantages of better protection, and of being more easily heated from the smaller area of [Pg 138] glass exposed to radiation. These designs are of houses of a cheap class, such as might be erected for merely experimental purposes.

Fig. 58.—*Perspective.*

[Pg 139]

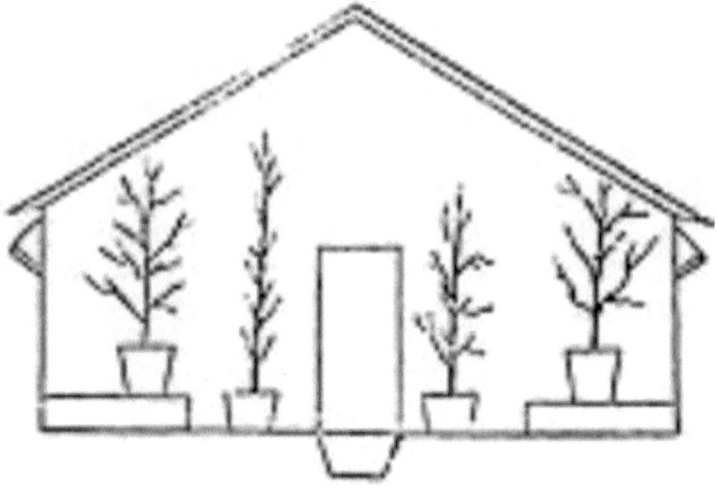

Fig. 59.—*Interior View.*

We consider the successful cultivation of Orchard fruit under glass, to be a fact so well settled, that we should advise substantial structures to be erected at the outset. Some of our numerous designs for graperies, both of the curvilinear and straight roofed form, would, with slight alteration in adding to the means of ventilation, be well adapted to this purpose. This is especially the case with designs numbered 7, 8, and 14.

www.ingramcontent.com/pod-product-compliance
Lightning Source LLC
Chambersburg PA
CBHW031442210526
45464CB00005B/2302